The Many Faces of
Venus

Ev Cochrane

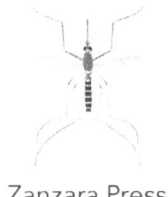

Zanzara Press

Zanzara Press
918 5TH ST
Ames, IA 50010
USA
zanzarapress.com
editor@zanzarapress.com

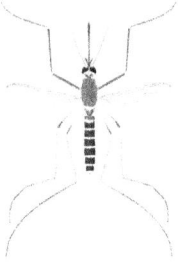

Zanzara Press

THE MANY FACES OF VENUS
2022 © Ev Cochrane, previously published in 2001 by Aeon Press

All rights reserved. No part of this work covered by the copyright hereon may be reproduced or used in any form or by any means–graphic, electronic, or mechanical, including photocopying, recording, taping, or information storage and retrieval systems–without written permission of the publisher. Neither the author nor the publisher make any representation, express or implied, with regard to the accuracy of the information contained in this book and cannot accept any legal responsibility or liability for any errors or omissions that may be made.

ISBN: 978-1-941892-53-4

Book layout and design by polytekton.com

Table of Contents

Preface to the Second Edition 5
Introduction 7

1. The Female Star 12
2. Inanna: Warrior-Goddess Extraordinaire 18
    The Sacred Marriage Rite 22
3. A Marriage Made in Heaven 25
4. The Many Loves of Aphrodite 34
5. Aphrodite and Phaon 44
6. Aphrodite and Adonis 50
7. Aphrodite: Star of Lamentation 56
8. The Venus-Comet 61
    Hair-Star 61
    Tailed-Star 64
    Bearded-Star 65
    Torch-Star 66
    Dragon-Star 68
    "Smoking star" 70
    Summary 71
9. Aphrodite: Warrior Goddess 73
    Kali 78
10. The "Witch-Star" 83
    Holda 84
    Hecate 86
    Nahtfare 87

11. Sovereignty as Hag:	89
A Case Study in Mythological Analysis	89
12. Hathor: The Egyptian Venus	97
The Eye of Horus	97
13. The Latin Goddess Venus	112
On the Origins of Magic	113
Venus as Binder	114
Venus in Ancient Mesopotamia	116
The Uraeus-Goddess	117
Inanna/Venus and the Crown of Kingship	118
The Shen Bond	121
Venus and the Lead-Rope of Heaven	123
Summary	125
Conclusion	127
Appendices	
Why Study Myth?	132
Suns and Planets in Prehistoric Rock Art	137
Solar Imagery in Ancient Mesopotamia	142
Bibliography	149

Preface to the Second Edition

"The inertia of the human mind and its resistance to innovation are most clearly demonstrated not, as one might expect, by the ignorant mass—which is easily swayed once its imagination is caught—but by professionals with a vested interest in tradition and in the monopoly of learning. Innovation is a twofold threat to academic mediocrities: it endangers their oracular authority, and it evokes the deeper fear that their whole, laboriously constructed intellectual edifice might collapse. The academic backwoodsmen have been the curse of genius from Aristarchus to Darwin and Freud; they stretch, a solid and hostile phalanx of pedantic mediocrities, across the centuries." [1]

The Many Faces of Venus was originally published in 2001 and, at the time, represented the most complete compendium of Venus-lore ever assembled. In the preface to that first edition I pointed out that it was still a work in progress and much remained to be done. The mythology attached to the planet Venus is vast in scope and the task of properly cataloguing and analyzing it all is not only daunting, it will likely require multiple generations of specialists in assorted disciplines to sort it all out. That said, the basic thesis of *Many Faces* remains intact and gains in credence with each passing day. Briefly stated: It is our claim that, if the testimony of the ancient skywatchers is to be believed, the Earth was a participant in a series of *recent* interstellar cataclysms of a virtually unimaginable nature—cataclysms that were devastating in nature and traumatic in psychological impact. It can be shown, moreover, that such planetary events had a formative influence on the primary institutions of early cultures and thus their influence continues to be felt to this very day.

---

[1] A. Koestler, *The Sleepwalkers* (New York, 1959), p. 427.

In this second edition I have done a fair amount of editing. At various points I have rearranged the chapters, eliminating some while substituting others in an effort to clarify and bolster the argument, which I trust is now more focused and cogent. I have also updated the footnotes and references where new findings bear on the argument. It is hoped that future editions will allow for including a generous portion of the rich iconographical tradition associated with the Venus-goddess. Alas, the budget of a lone scholar toiling away in obscurity does not allow for such luxuries at the present time, desirable as they may be.

Introduction

"If we look at the physical universe the way astronomers do, we may never know anything about it. The recent U.S. planetary probes revealed a shocking paucity of real knowledge about the contents of the cosmos."[2]

"One thing is certain, the idea that planets can change their orbits dramatically is here to stay."[3]

The slow and steady movement of the respective planets about the sun is frequently held out as a sign of the clock-like regularity and order which distinguishes the present solar system. Yet it can be shown that this apparent regularity represents a relatively recent development. As we will document in the pages to follow, the ancient skywatchers describe a radically different solar system. If we are to believe their collective testimony, recorded in sacred traditions and artworks from every corner of the globe, Venus only recently moved on a different orbit, cavorting with Mars and raining fire from heaven. Is it perchance possible that modern astronomers, in neglecting the manifold traditions surrounding these two planets, have overlooked a vital clue to the recent history of the solar system? I, for one, believe this to be the case.

The planet Venus has long fascinated terrestrial stargazers. Already at the dawn of recorded history, Sumerian priests composed hymns in honor of the planet which they worshipped as the warrior-goddess Inanna:

---

[2] G. Verschuur, "Revising the Truth," *Science Digest* (August, 1981), p. 29.
[3] Dr. Renu Malhotra of the Lunar and Planetary Institute of Houston as quoted in the *Sunday Telegraph* (October 10, 1999), p. 19.

"I shall greet her who descends from above...I shall greet the great lady of heaven, Inana! I shall greet the holy torch who fills the heavens, the light, Inana, her who shines like the daylight, the great lady of heaven, Inana! I shall greet the Mistress, the most awesome lady among the Anuna gods; the respected one who fills heaven and earth with her huge brilliance...Her descending is that of a warrior."[4]

As the earliest historical testimony regarding Venus, the Sumerian Inanna texts are of paramount importance in reconstructing the ancient conceptions surrounding our Sister planet. That said, many of the statements from these texts are difficult to reconcile with Venus's current appearance and customary behavior. In what sense can it be said that Venus "fills heaven with her huge brilliance" and/or descends as a warrior? The following passage, which describes the planet-goddess as dominating the skies and raining fire and destruction, is representative in this regard:

"Inanna, who pours down rain over all the lands, over all the people, loud-thundering storm. Hierodule, who makes heavens tremble, who makes the earth quake, Who can soothe your heart? You who pour down firebrands over the earthly orb, who flash like lightning over the highland...Whose cry reaches heaven and earth, whose roar is all-destructive...Your angry heart is a terrifying flood-wave."[5]

Scholars investigating the ancient hymns devoted to Inanna/Venus have typically viewed the vivid catastrophic imagery as the product of figurative language and poetic license. Yet it can be shown that the Sumerian imagery has precise parallels around the globe, in the New World as well as the Old.

In Mesoamerica, for example, where the observation and veneration of the celestial bodies amounted to a collective obsession, the appearance of Venus was an occasion marked by dread and apocalyptic paranoia for the Aztecs and Maya alike. Bernardino de Sahagún, a Franciscan friar writing in the 16th century, chronicled the Aztec angst towards Venus as follows:

"And when it [Venus] newly emerged, much fear came over them; all were frightened. Everywhere the outlets and openings of [houses] were closed up. It was said that perchance [the light] might bring a cause of sickness, something evil when it came to emerge."[6]

---

[4] Lines 1-18 as quoted from "A šir-namursaga to Inana for Iddin-Dagan," in J. Black et al, *The Literature of Ancient Sumer* (Oxford, 2004), p. 263. See also D. Reisman, "Iddin-Dagan's Sacred Marriage Hymn," *Journal of Cuneiform Studies* 25 (1973), pp. 186-191.
[5] S. Kramer, *From the Poetry of Sumer* (Berkeley, 1979), p. 89.
[6] B. de Sahagún, *Florentine Codex. General History of the Things of New Spain, Book 7* (Santa Fe, 1953), p. 12.

In an effort to stave off the planet's wrath, the various cultures of Mesoamerica offered it human sacrifices on a scale which aroused horror and revulsion amongst early chroniclers. What is there about Venus that could have inspired such rituals, conducted amidst an atmosphere of solemn purpose? Venus's present appearance, needless to say, would never evoke mass panic or vivid tales of impending doom and world destruction. And yet the sense of dread apparent in de Sahagún's account mirrors the Sumerian skywatchers' view of Inanna/Venus from over three thousand years earlier: "Agitation, terror, fear, splendour, awe-inspiring sheen are yours, Inanna."[7]

In this book we will seek to discover the natural history and logical rationale behind the ancient traditions attached to Venus. To anticipate our conclusion: Venus was associated with dire portents and fears of apocalyptic disaster for a very simple reason—It was a primary player in spectacular cataclysms involving the Earth in relatively recent times, well within the memory of humankind.

The implications of this finding, if true, are at once revolutionary and far-reaching. In addition to necessitating a drastic revision in our understanding of the wellsprings of ancient myth and religion, the central tenets of modern astronomy and a host of allied sciences would be called into question as well. With stakes this high, it is imperative that we gain additional insight into the ancient traditions surrounding Venus.

The ancients' obsession with the planet Venus stands in marked contrast to the relative indifference currently accorded that planet. Who among us could even point out the Evening Star on any given night? Would anyone in their right mind be inclined to view Venus as a terrifying agent of destruction and impending doom?

David Grinspoon, a NASA astronomer and the author of a very entertaining history of Venus observation, offered the following summary of humankind's preoccupation with our so-called Twin, one which might well be regarded as representative of conventional opinion:

"Venus must always have seemed a unique, animated entity. For our ancestors the details of the complex movements of Venus served as important harbingers of war and peace, feast and famine, pestilence and health. They learned to watch every nuance for the clues they could wrest of what nature had in store. They watched carefully, obsessively, through skies not yet dimmed by industrial haze and city lights, and

---

[7] Å. Sjöberg, "in-nin šà-gur$_4$-ra. A Hymn to the Goddess Inanna...," *Zeitschrift für Assyriologie* 65 (1976), p. 195.

they learned to predict accurately, for years and decades to come, the rising, setting, dimming, brightening, and looping of Venus."[8]

Faced with the abundant evidence documenting Venus's prominent role in ancient culture, Grinspoon, like countless others before him, takes it for granted that it is only natural that native skywatchers would look to that particular planet for omens of things to come. But why should this be, since there is neither an inherent nor logical relation between Venus and the phenomena mentioned by him—war, pestilence, famine, etc.?[9] Indeed, it stands to reason that any skywatcher worth his salt would soon discover that there was precious little to be learned about such matters from the patient observation of Venus. That is, of course, if we are to believe the conventional version of Venus's history, which holds that its orbit and appearance has not changed in any significant manner for many millions of years.

In recent years, modern astronomers have been surprised by space discoveries at every turn. For the first several centuries of telescope observation it was commonly believed that Venus was home to beings like ourselves, replete with a thriving civilization.[10] Until the midpoint of the present century it was still thought possible that Venus might be "Earth-like" in its features, with a tropical climate, vast oceans and swamps teeming with various forms of life.[11] Yet all such geocentric scenarios were to receive a severe jolt in 1962 when, courtesy of Mariner 2, Venus was revealed to be an incredibly inhospitable place, with surface temperatures in excess of 900 degrees Fahrenheit. Under such conditions, oceans (of water, that is) are quite out of the question and life, as we know it, virtually unthinkable.[12]

Despite much progress, the recent history of Venus investigation reveals a vast graveyard of shoddy observations, false deductions, and discarded hypotheses.[13] While leading astronomers expected the Venusian clouds to be composed of water, it turned out that they have precious little water and

---

[8] D. Grinspoon, *Venus Revealed* (New York, 1997), p. 17.

[9] Polly Schaafsma, E. Krupp, S. Milbrath, M. Mathiowetz, and R. Hall, "The Role of Venus in the Cosmologies of Mesoamerica, West Mexico, the American Southwest, and Southeast," p. 1 concede the point: "It is clear that in real world terms Venus and its wanderings can have no physical impact on human societies." Note: This is a white paper distributed on Academia.edu.

[10] D. Grinspoon, *op. cit.*, pp. 31-39.

[11] Of Ray Bradbury's story "The Long Rain," which describes Venus as an "Earthlike, rain-soaked, heavily vegetated jungle world," Grinspoon, *op. cit.*, p. 31 says that it was "consistent with common scientific beliefs of the day"—i.e., 1951.

[12] R. Malcuit, *The Twin Sister Planets Venus and Earth* (London, 2015), pp. 1-8.

[13] For a survey, see D. Grinspoon, *op. cit.*, pp. 40-45.

plenty of concentrated sulfuric acid. Where astronomers "observed" luxuriant Venusian vegetation in full bloom,[14] modern space probes discovered a barren, desiccated wasteland. And so it goes. Indeed, if the truth be told, the Mariner, Magellan, and Pioneer missions have forced astronomers to radically revise their previous assessments as to Venus's origin, atmosphere, and geological history. On virtually every fundamental feature of the Venusian landscape and atmosphere, the astronomers' theoretical expectations have been proven wrong time and again. And wildly wrong at that. Given this dismal track record, there would appear to be some justification for maintaining a healthy skepticism with respect to NASA's current "best guesses" as to what is most likely regarding Venus's recent history and fundamental nature. Indeed, there are good reasons for believing that other—even more radical—revisions in our understanding of Venus and its recent history will soon be in order.

---

[14] *Ibid.*, p. 51.

1. The Female Star

"What is it that gives Venus this particularly exalted place in mythology?"[15]

"Mythology occupies a vital place in the worldview of most people, particularly among those without written traditions. Mythology was essentially indistinguishable from what modern people might call 'reality' or 'history.' Often tied to religious belief, mythology can be a window into the past and the means by which people make sense of the present."[16]

"These stories, however, had a point. They meant something. As we revisit them, we should also be trying to understand exactly how they once made the sky comprehensible. This requires mindful study of their content. Detailed analysis of the content of any celestial myth in turn should be directed to the revelation of the actual function of that myth."[17]

A survey of ancient Venus-lore reveals an hitherto unsuspected wealth of endlessly recurring mythological motifs: the planet as mother goddess; the planet as agent of war, death, and destruction; the planet as witch-like hag; the planet as paramour of Mars; and countless others. A systematic analysis of the manifold mythological themes attested around the globe, in turn, allows for the reconstruction of an archetypal Venus myth which, in a very real sense, constitutes humankind's collective memory of that planet's recent history.

While many of our claims with respect to the mythological themes associated with Venus will sound fantastic when enumerated in advance of proper documentation, its proverbial association with the fairer sex is commonly acknowledged and readily demonstrable. Thus, in his recent book on the planet,

---

[15] G. Santillana, "Prologue to Parmenides," in *Reflections on Men and Ideas* (Cambridge, 1970), p. 116.
[16] B. Pritzker, *A Native American Encyclopedia* (Oxford, 2000), p. xi.
[17] E. Krupp, "Sky Tales and Why We Tell Them," in H. Selin ed., *Astronomy Across Cultures* (Dordrecht, 2000), p. 20.

Peter Cattermole noted that "a female association is almost universal."[18] And so it is, despite occasional statements to the contrary.[19]

The early Hebrews, for example, knew Venus as *Kokabat*, a name which translates as "she-star."[20] This name finds a cognate in the Syrian *Kawkabta*, also attested as a female star.[21] Among the Arabic peoples of Northern Syria and the Mesopotamian desert, Venus was known as *al-'Uzza* and conceived as an invincible warrior.[22] She, too, was deemed to be of female form: "In sources from the fifth century AD she is identified with Aphrodite by an anonymous Syrian historian; with *Kaukabta*, 'the female star;' with Balthi, by Isaac of Antioch; and finally with Lucifer, the morning star, by Jerome."[23]

In ancient Persia, Venus was identified with the voluptuous goddess Anahita, the latter being conceptualized as a formidable warrior and agent of fertility.[24] A vestigial remnant of these ancient traditions is apparent in the *Koran*, wherein one verse describes the transfiguration of a young woman into the beautiful star Zohra, the Arabic name for Venus.[25] As has been pointed out, variant traditions name this young woman *Anahid*.[26]

Similar conceptions are to be found among the indigenous peoples of Siberia, where the Yakut knew the planet by the name *Solbon*, envisaged as a beautiful girl.[27] A legend first recorded in the last century reads as follows:

---

[18] P. Cattermole, *Venus: The Geological Story* (Baltimore, 1994), p. 1.

[19] Following the lead of Anthony Aveni, Grinspoon, *op. cit.*, p. 24, writes that: "It is simply not true that a female association is general." In support of this statement, Grinspoon points to Quetzalcoatl and Tlahuizcalpantecuhtli as classic male Venus deities. Here it can be shown that most of the male deities which various scholars have hitherto identified with Venus are actually Martian in origin. See E. Cochrane, *Starf*cker* (Ames, 2006), pp. 94-139.

[20] R. Stieglitz, "The Hebrew Names of the Seven Planets," *Journal of Near Eastern Studies* 40:2 (1981), pp. 135-136. See L. Bobrova & A. Militarev, "From Mesopotamia to Greece: to the Origin of Semitic and Greek Star Names," in H. Galter ed., *Die Rolle der Astronomie in den Kulturen Mesopotamiens* (Graz, 1993), p. 315.

[21] *Ibid.*

[22] J. Healey, The Religion of the Nabataeans (Leiden, 2001), p. 117: "That al-'Uzza was both in the Arabian and in the northern context a planetary deity representing the morning star, Venus, is very clear from a wide range of sources."

[23] W. Heimpel, "A Catalog of Near Eastern Venus Deities," *Syro-Mesopotamian Studies* 4:3 (1982), p. 19.

[24] A. Carnoy, "Iranian Mythology," in L. Gray ed., *The Mythology of All Races* (Boston, 1917), pp. 279-280. See *Yast* 5.85 for Anahita's astral aspect. See also M. Saadi-nejad, *Anahita* (London, 2021), p. 13.

[25] *Koran* 2.96. See also W. Eilers, *Sinn und Herkunft der Planetennamen* (München, 1976), p. 55.

[26] J. Puhvel, *Comparative Mythology* (Baltimore, 1989), p. 104.

[27] L. Mándoki, "Two Asiatic Sidereal Names," in V. Dioszegi ed., *Popular Beliefs and Folk-*

> *"She is the bride and sweetheart of Satan's son—ürgel...When these two stars come close to one another, it is a bad omen; their eager quivering, their discontinuous panting cause great disasters: storms, blizzards, gales. When they unite, fathomdeep snow will fall even in the summer, and all living beings, animals and trees will perish..."*[28]

The Yakut believed that the appearance of Venus heralded ominous portents. Witness the following report:

> *"It is said to be 'the daughter of the Devil and to have had a tail in the early days.' If it approaches the earth, it means destruction, storm, and frost, even in the summer; 'Saint Leontius, however, blessed her and thus her tail disappeared.'"*[29]

Viewed in isolation, such traditions can only seem the stuff of fantasy. Yet when viewed with a critical eye in the light of comparable traditions from around the globe, the suspicion arises that something more than mere fantasy is at work here.

The malevolent nature accorded the Cytherean planet as evinced by the aforementioned Aztec and Yakut traditions will prove to be a recurring theme throughout the course of this book. Among the Samoan Islanders, it was said that Venus (*Tapuitea*) was a primeval Fijian queen who, upon going berserk, suddenly sprouted horns from her head and engaged in cannibalistic practices.[30] Shortly thereafter, the queen was translated to the heavens where, as Venus, she continues to presage the death of a kings and nobles.

Venus was widely regarded as female in the New World well. Among the Zinacanteco Indians of Mesoamerica, Venus was conceptualized as a girl sweeping the path of the sun.[31] Further south the Inca knew Venus as a lovely woman by the name of *Chasca*. An anonymous Jesuit of the 17th century described the planet-goddess as especially devoted to women and princesses:

> *"[Of Venus] they said that she was a goddess of young maidens and princesses, and originator of the flowers of the fields, and mistress of dawn and twilight; and it was she who threw dew onto the earth when she shook her hair, and they thus called her Chasca [the Disheveled One]."*[32]

---

lore *Traditions in Siberia* (Bloomington, 1968), p. 489.

[28] *Ibid.*, p. 489.

[29] *Ibid.*, p. 489.

[30] R. Williamson, *Religious Beliefs and Cosmic Beliefs of Central Polynesia, Vol 1* (Cambridge, 1933), p. 128.

[31] See J. Sosa, "Maya Concepts of Astronomical Order," in G. Gossen ed., *Symbol and Meaning Beyond the Closed Community* (Albany, 1986), p. 189. See also the tale quoted by E. Vogt, *Zinacantan* (Cambridge, 1969), pp. 316-318.

[32] *De las costumbres antiquas de los naturales del Piru* (Madrid, 1879), as translated by Jan

For the indigenous tribes of the Amazonian rain forest, a popular story conceptualized Venus as a woman who descends from heaven in order to find a mortal lover. The anthropologist Johannes Wilbert summarized these stories as follows:

"Planets appear to be women who, like Morning Star, may descend to earth and become the consorts of men. Venus is the Star Woman who takes pity on a rejected man. She appears in several beautiful text versions of the collection which resemble in form and content the celestial bride tales of other Chaco mythologies, as well as their counterparts in the Ge tradition of eastern Brazil. While associated with her terrestrial companion, the Morning Star of Chorote mythology repeatedly demonstrates her power to destroy human adversaries and their fields."[33]

The Australian Aborigines are renowned for their skywatching habits and abilities and, not surprisingly, the planets play a conspicuous role in their sacred tales, rituals, and artwork.[34] There, too, the planet Venus was viewed as a female being:

"The Morning Star was also an important sign to the Aborigines who arose at early dawn to begin their hunting. It, too, was personified and frequently associated with death. Arnhem Land legends identify the home of the morning star, Barnumbir, as Bralgu, the Island of the Dead. Barnumbir was so afraid of drowning that she could be persuaded to light her friends across the sea at night only if she were held on a long string by two old women, who at dawn would pull her back to shore and keep her during the day in a basket. In Arnhem Land, because of this connection, the morning star ceremony is an important part of the ritual for the Dead. Barnumbir is represented by a totem stick to the top of which is bound a cluster of white feathers or down, denoting the star, and long strings ending in smaller bunches of feathers to suggest the rays. When a person dies, his/her spirit is believed to be conducted by the star to Bralgu, its last resting place."[35]

The intimate connection of Venus with the land of the dead, apparent here, is archetypal in nature and attested around the world.[36]

Sammer, "The Cosmology of Tawantinsuyu," *Kronos* 9:2 (1984), p. 25.

[33] J. Wilbert & K. Simoneau, *Folk Literature of the Chorote Indians* (Los Angeles, 1985), p. 7.

[34] H. Cairns, "Aboriginal sky-mapping," in C. Ruggles ed., *Archaeoastronomy in the 1990's* (Loughborough, 1993), p. 136.

[35] R. Haynes, "Aboriginal Astronomy," *Australian Journal of Astronomy* 4:3 (1992), p. 134.

[36] E. Cochrane, *The Many Faces of Venus* (Ames, 2001), pp. 167-179. See also E. Krupp,

A widespread belief finds the planet Venus being viewed as the daughter of the ancient sun-god. The following account, from the Tsimshian Indians of North America, offers a representative example of this theme:

"The sky is a beautiful open country. It is reached through the hole in the sky, which opens and closes...The sky may also be reached by means of a ladder which extends from the mountains to the sky...After reaching the sky, the visitor finds himself on a trail which leads to the house of the Sun chief. In this house the Sun lives with his daughter... The Sun's daughter is the Evening Star." [37]

The same general idea is found among the Desana Indians of the Amazonian rain forest, who knew Venus as "Daughter of the Sun."[38] According to the Desana, Venus was the victim of incestuous advances at the hands of her father. Strangely enough, however, it is reported that she continued to live with him as his wife.

Among various Slavic peoples, Venus was known by the name of Danica, "day-star," and viewed as the sister of the sun.[39] This same planet-goddess was elsewhere recalled as a great king's daughter.[40]

A number of hypotheses have been advanced in order to explain the Venus-as-female theme, none of which will bear scrutiny. The astronomer Patrick Moore offered the following suggestion: "A female association is in fact general, except in India; this is natural enough, since to the unaided eye Venus is the loveliest of the planets."[41] Here's a male chauvinist answer if ever there was one. Why should a beautiful planet be viewed as female rather than male?

Others have sought an answer in peculiarities in the planet's periods of visibility. David Grinspoon, for example, offered the following argument:

"Another reason to worship Venus and find significance in her movements is that there are numerous connections between the timing of aspects of her motions and timescales of natural interest to humans. Perhaps most strikingly, the approximate 260-day length of a Venus appearance in the morning or evening coincides closely with the average length of the human gestation period...A knowledge of the close coincidence between the cycles of Venus and human pregnancy

---

Beyond the Blue Horizon (New York, 1991), p. 197ff.

[37] F. Boas, "Tsimshian Mythology," *ARBAE* 31 (1916), pp. 453-454.

[38] G. Reichel-Dolmatoff, *Amazonian Cosmos* (Chicago, 1971), pp. 28-29, 71.

[39] William Gibbon, *Popular Star Names Among the Slavic Speaking Peoples*, PhD dissertation, University of Pennsylvania, 1960.

[40] *Ibid.*, p. 180.

[41] D. Grinspoon, *op. cit.*, p. 24, quoting Patrick Moore's *The Planet Venus* (1959).

may have contributed to the persistent, but nonexclusive, Western attribution of female characteristics to Venus."[42]

This explanation, while more logically consistent than that of Moore, can hardly be given credence, for it implies that primitive cultures around the world were aware of the 260-day appearance interval of Venus and noticed its near-synchronism with the human gestation period. Yet there is little evidence for this claim and plenty against it. While the 260-day period of Venus was certainly known to the Maya and was likely known to the ancient Babylonians of the first millennium BCE, it can hardly be believed that the Australian Aborigines or the Siberian Yakut—both of whom regarded Venus as a female—were aware of this period.

The logical problems evident in these explanations underscore the fact that it is very difficult to discern anything in the present appearance of Venus which would justify its traditional femininity. Indeed, it is our opinion that nothing in the current appearance and/or behavior of Venus will explain this widespread and persistent theme—nor little else in the ancient mythology surrounding that planet, for that matter. Yet before we attempt an explanation of this theme, it is necessary to briefly review the cult of the Sumerian Inanna, where the feminine gender of the planet Venus is front and center.

---

[42] *Ibid.*, p. 18. E. Krupp, *Beyond the Blue Horizon* (New York, 1991), p. 184 offered a similar explanation.

2. Inanna: Warrior-Goddess Extraordinaire

> *"The astral identity of Inanna was not the result of late, learned speculations, but indeed a very old and fundamental aspect of the goddess, with roots going back to prehistoric times."*[43]
>
> *"It is remarkable that the ancient Mesopotamians put so much stress on the empirical observation of the Venus's orbit [sic]. This proves two things: firstly, that empiricism is an important ingredient of Mesopotamians' theological speculations, and secondly, that the reconstruction of this sort of empiricism is indispensable for any understanding of the Mesopotamian 'belief system'."*[44]
>
> *"If we survey the whole of the evidence on this subject…we may conclude that a great Mother Goddess, the personification of all the reproductive energies of nature, was worshipped under different names but with a substantial similarity of myth and ritual by many peoples of Western Asia; that associated with her was a lover, or rather series of lovers, divine yet mortal, with whom she mated year by year, their commerce being deemed essential to the propagation of animals and plants, each in their several kind; and further, that the fabulous union of the divine pair was simulated and, as it were, multiplied on earth by the real, though temporary, union of the human sexes at the sanctuary of the goddess for the sake of thereby ensuring the fruitfulness of the ground and the increase of man and beast."*[45]

One of the greatest goddesses of the ancient world was the Sumerian Inanna who, as is well known, was explicitly identified with the planet Venus already at the dawn of history. The cult of Inanna, upon being assimilated with that of the Semitic goddess Ishtar, would dominate the religious landscape of Mesopotamia for well over two thousand years. As our earliest historical testimony documenting the observation and worship of the planet Venus, the

---

[43] P. Beaulieu, *The Pantheon of Uruk During the Neo-Babylonian Period* (Leiden, 2003), p. 104.

[44] G. Selz, "Mythological Narratives and Their Backgrounds," in J. Gießauf ed., *Zwischen Karawane und Orientexpress* (Münster, 2017), p. 279.

[45] J. Frazer, *Adonis, Attis, Osiris* (New Hyde Park, 1961), p. 39.

literature surrounding Inanna must figure prominently in any discussion of that planet's role in ancient myth and religion.

Inanna's cult is ubiquitous in the earliest religious temples yet excavated in Mesopotamia.[46] At Uruk, the oldest urban site in the entire Near East, offerings to Inanna/Venus far outnumber those of any other deity.[47] In strata conventionally dated to ca. 3000 BCE, Inanna is already associated with various symbols that would become conspicuous in her later cult (the eight-pointed star and rosette, for example).

Writing first developed in Sumer during this same period (i.e., Uruk IV/III), later spreading to Egypt. In the most archaic texts from Uruk—conventionally attributed to roughly 3300 BCE[48]—Inanna's name was written with a pictographic sign known as MUŠ$_3$, commonly interpreted as a gate-post with streamer (see figure 1).[49] Elizabeth Van Buren observed that a single gate-post with streamer "is always used on pictographic tablets to symbolize the goddess," and that it frequently appears together with the star-sign.[50] Although most examples of this sign have been found at the archaic E-anna precinct at Uruk, it also occurs elsewhere in Mesopotamia during the same period.[51]

Figure 1

In the earliest period (Uruk IV), the sign typically appears without the divine determinative, although exceptions do occur. In the subsequent period,

---

[46] According to Wolfgang Heimpel, "A Catalog of Near Eastern Venus Deities," *Syro-Mesopotamian Studies* 4:3 (1982), p. 12 the identification of Inanna and Venus was first made in prehistoric times and is apparent "in all historical periods."
[47] K. Szarzynska, "Offerings for the goddess Inana," *Revue d'assyriologie et d'archéologie orientale* 87 (1993), p. 7.
[48] J. Glassner, *The Invention of Cuneiform* (Baltimore, 2003), p. 45.
[49] C. Wilcke, "Inanna/Ishtar," *Reallexikon der Assyriologie* 5 (Berlin, 1976-1980), pp. 74-75. For the various examples of this sign in archaic texts, see M. Green & H. Nissen, *Zeichenliste der archaischen Texte aus Uruk* (Berlin, 1987), p. 248.
[50] E. Van Buren, *Symbols of the Gods in Mesopotamian Art* (Rome, 1945), p. 43.
[51] K. Szarzynska, "Some of the Oldest Cult Symbols in Archaic Uruk," *Jaarbericht ex Oriente Lux* 30 (1987-88), p. 10.

the MUŠ3-sign is usually preceded by a divine determinative—a star-like object closely resembling a modern asterisk. Inasmuch as the pictographic determinative for "god" features an eight-pointed star, it stands to reason that Inanna was already identified with a celestial body during the archaic Uruk period. Krystyna Szarzynska, a leading authority on the archaic Uruk script, has expressed a similar view: "In the most archaic period the determinative dingir was associated with astral deities only."[52]

The earliest literary texts from Mesopotamia date from the Early Old Babylonian period (ca. 2000-1800 BCE).[53] The corpus of hymns allegedly composed by Sargon's daughter Enheduanna is representative of this period and literary genre. The hymn nin-me-šar-ra, generally known as "The Exaltation of Inanna," rarely mentions the goddess by name; instead Inanna is invoked through a series of epithets such as "great queen of queens"[54] or "hierodule of An."[55] As the planet Venus, Inanna is celebrated as "senior queen of the heavenly foundations and zenith."[56] Here, as elsewhere, Inanna personifies Venus as the "beloved bride" of Dumuzi.[57]

Other epithets of the goddess are more difficult to explain by reference to the planet Venus and, as a result, scholars have been inclined to interpret them by reference to figurative language. Included here are those epithets describing Inanna as a fire-spewing dragon and agent of the storm:

"Like a dragon you have deposited venom on the land, When you roar at the earth like Thunder, no vegetation can stand up to you. A flood descending from its mountain, Oh foremost one, you are the Inanna of heaven and earth! Raining the fanned fire down upon the nation...When mankind comes before you In fear and trembling at your tempestuous radiance."[58]

Inanna's prowess as a warrior is a recurring point of emphasis in the Sumerian texts. The hymn in-nin šà-gur$_4$-ra describes the planet-goddess as a terrifying warrior "covered in storm and flood."[59] The following passage is representative:

---

[52] Personal communication, March 22, 1997. It was Szarzynska's opinion that the veneration of astral deities reaches back to the proto-Sumerian period, perhaps earlier.
[53] J. Hayes, *A Manual of Sumerian Grammar and Texts* (Malibu, 2000), p. 394.
[54] See W. Hallo & J. van Dijk, *The Exaltation of Inanna* (New Haven, 1968), p. 23.
[55] *Ibid.*, p. 15.
[56] *Ibid.*, p. 29.
[57] *Ibid.*, p. 29.
[58] *Ibid.*, pp. 15-17.
[59] A. Sjöberg, "in-nin šà-gur$_4$-ra. A Hymn to the Goddess Inanna…," *Zeitschrift für Assyriologie* 65 (1976), p. 181.

> *"Goddess of the fearsome divine powers, clad in terror...Inana... drenched in blood, rushing around in great battles...covered in storm and flood, great lady Inana...In heaven and on earth you roar like a lion and devastate the people."*[60]

In another passage from the same hymn the warrior-goddess is represented as threatening the gods in heaven:

> *"She is a huge neckstock clamping down on the gods of the land, Her radiance covers the great mountain, silences the road, The gods of the land are panic-stricken by her heavy roar, At her uproar the Anunna-gods tremble like a solitary reed, At her shrieking they hide all together."*[61]

A recurring epithet of the planet-goddess in these early texts—*an al-dúb-ba ki sìg-ga*, "[she] who shakes the sky and makes the earth tremble"[62]—commemorates her destructive nature. In these early hymns Inanna/Venus is depicted as an awe-inspiring numinous power, to be feared as well as propitiated. The following passage is representative in this regard: "Agitation, terror, fear, splendour, awe-inspiring sheen are yours, Inanna!"[63] Once again it is necessary to ask: What is there about Venus's appearance or behavior that would evoke terror to the point of dread?

The apparent incongruity between the goddess as planet and the goddess as warrior, together with various other peculiarities of her cult, has led scholars to suppose that Inanna's personality represents a jumble of contradictions. Samuel Kramer, for example, spoke of the "contrasting strands in Inanna's multi-faceted character."[64] Thorkild Jacobsen, similarly, observed that "the offices attributed to her show little unity or pattern."[65]

Upon failing to discern a recognizable pattern behind the manifold aspects of the planet-goddess, Sumerologists have proposed that a number of originally separate goddesses coalesced to form the multivalent Inanna. This is the view defended by Jacobsen: "Actually Inanna has a good many more aspects than those which characterize her in her relations with Dumuzi, so many different ones in fact that one is inclined to wonder

---

[60] Lines 1-7 from "Inana and Ebih," as translated in J. Black et al, *The Literature of Ancient Sumer* (Oxford, 2004), p. 334.

[61] A. Sjöberg, "in-nin šà-gur₄-ra. A Hymn to the Goddess Inanna...," *Zeitschrift für Assyriologie* 65 (1976), p. 179.

[62] *Ibid.*

[63] *Ibid.*, line 161.

[64] Quoted from R. Harris, "Inanna-Ishtar as Paradox and a Coincidence of Opposites," *History of Religions* 30:3 (1991), p. 262.

[65] T. Jacobsen, "Mesopotamian Religion: An Overview," in M. Eliade ed., *The Encyclopedia of Religion* (New York, 1987), p. 459.

whether several, originally different deities have not here coalesced into one, the many-faceted goddess Inanna."[66]

It is our opinion that such views of Inanna's origins and fundamental nature are hopelessly wrongheaded. While it is impossible to rule out the possibility that syncretism played a minor role in the goddess's cult, the fact remains that the vast majority of Inanna's attributes make perfect sense in light of ancient conceptions associated with the planet Venus. Even the planet-goddess's warrior-like persona—the bane of scholarly attempts to find a rational explanation of Inanna's original nature—finds a ready explanation from this vantage point, Venus being widely regarded as a warrior, in the ancient Near East as well as in China and the New World.[67]

The Sacred Marriage Rite

"The Babylonian paradigm for love and marriage was the relationship between Inanna and Dumuzi."[68]

"The nature of a ritual of kingship known as the 'Sacred Marriage' has long puzzled scholars."[69]

One of the most important celebrations in ancient Mesopotamia was the so-called sacred marriage rite, purporting to reenact the sexual union of Inanna with Dumuzi at the time of Beginning.[70] Of untold antiquity—a vase recovered from the Protoliterate period at Uruk (ca. late 4th millennium BCE) is thought to depict the marriage of Inanna and Dumuzi[71]—the ritual appears to have died out after the Old Babylonian period.[72]

Early texts report that the performance was believed to stimulate the growth

---

[66] *The Treasures of Darkness* (New Haven, 1976), p. 135.

[67] J. Major, *Heaven and Earth in Early Han Thought* (Buffalo, 1993), p. 76.

[68] W. Heimpel, "Mythologie, A. I," *Reallexikon der Assyriologie, Vol. 8* (Berlin, 1993-1997), p. 547.

[69] K. McCaffrey, "The Sumerian Sacred Marriage: Texts and Images," in H. Crawford, *The Sumerian World* (New York, 2013), p. 227.

[70] E. van Buren, "The Sacred Marriage in Early Times in Mesopotamia: Part I," *Orientalia* 13 (1944), p. 1 states that "the very ancient rite of the sacred marriage was of the utmost importance, if not the essential and pivotal element of Babylonian religion."

[71] H. Frankfort, *The Art and Architecture of the Ancient Orient* (1954), pp. 25-27. See also G. Selz, "Five Divine Ladies," *NIN* 1 (2000), p. 31.

[72] R. Kutscher, "The Cult of Dumuzi/Tammuz," in J. Klein ed., *Bar-Ilan Studies in Assyriology* (New York, 1990), p. 41. Although references to a sacred marriage rite are to be found in the letters of Esarhaddon and Assurbanipal, human beings no longer take an active role in consummating the marriage of the goddess and her consort. See also D. Frayne, "Notes on The Sacred Marriage Rite," *Bibliotheca Orientalis* 42:1/2 (1985), cols. 11, 22.

of crops. In the rite in question a "flowered bed" or "garden" would be prepared whereupon the king would have intercourse with a woman representing Inanna.[73] Douglas Frayne offered the following summary of the rite:

"It is clear that the central purpose of the Sacred Marriage Rite was to promote fertility in the land. The rationale of the ceremony was that by a kind of sympathetic act involving the sexual union of the king, playing the role of the en [typically personifying Dumuzi] with a woman, generally referred to simply as Inanna, the crops would come up abundantly and both the animal and human populations would have the desire and fertility to ensure that they would multiply."[74]

Our most important source describing the rite is the so-called marriage hymn of Iddin-Dagan, the third king of the First Dynasty of Isin (ca. 1974-1954 BCE). The text begins by invoking Inanna as the planet Venus. Excerpts from the hymn follow:

"I shall greet her who descends from above…I shall greet the great lady of heaven, Inana! I shall greet the holy torch who fills the heavens, the light, Inana, her who shines like the daylight, the great lady of heaven, Inana! I shall greet the Mistress, the most awesome lady among the Anuna gods; the respected one who fills heaven and earth with her huge brilliance…Her descending is that of a warrior."[75]

Inanna is here compared to a shining torch whose "huge" brilliance is said to "fill heaven" and shine like the daylight, descriptors that are extremely difficult to reconcile with Venus's present modest luster.

In the ensuing lines of the hymn there are allusions to various offerings given to Inanna. After the goddess bathes herself, a flowered bed is set up for her and the king to share. Properly prepared, the king—in the guise of Dumuzi (addressed by the epithet Ama'ušumgalanna here)—approaches the bed:

"On New Year's day, the day of ritual, They set up a bed for my lady. They cleanse rushes with sweet-smelling cedar oil, They arrange them (the rushes) for my lady, for their (Inanna and the king) bed… My lady bathes (her) pure lap, She bathes for the lap of the king… The king approaches (her) pure lap proudly, Ama'ušumgalanna lies down beside her, He caresses her pure lap…She makes love with him on her bed, (She says) to Iddin-Dagan: 'You are surely my beloved.'…

---

[73] See here the discussion in D. Frayne, *op. cit.*, cols. 14, 21.
[74] *Ibid.*, col. 6.
[75] Lines 1-18 as quoted from "A *šir-namursaga* to Inana for Iddin-Dagan," in J. Black et al, *The Literature of Ancient Sumer* (Oxford, 2004), p. 263. See also D. Reisman, "Iddin-Dagan's Sacred Marriage Hymn," *Journal of Cuneiform Studies* 25 (1973), pp. 186-191.

The palace is festive, the king is joyous. The people spend the day in plenty. Ama'ušumgalanna stands in great joy. May he spend long life on the radiant throne!"[76]

In ancient Mesopotamia, as elsewhere, the ritual *hieros gamos* formed a prominent feature of New Year's celebrations. By all accounts it was a particularly joyous occasion, the consummation of the royal marriage being followed by a period of feasting and revelry:

"The glad news of the successful accomplishment of the long rite having been communicated to the people who had been waiting in anxious expectation to learn the issue, there was an outburst of exultation and thanksgiving, followed by a great feast of which all partook, the newly-wedded pair, the visiting divinities, the whole multitude who, in gratitude for the fertility which was now assured, raised jubilant hymns to the sound of the lyre, flutes and drums."[77]

At this point an obvious question presents itself: What is there about the planet Venus that could have inspired the peculiar details of the sacred marriage rite? For a possible answer to this question we turn to the New World, where strikingly similar conceptions appear among the Skidi Pawnee.

---

[76] D. Reisman, *op. cit.*, pp. 190-191.
[77] E. van Buren, *op. cit.*, p. 34.

3. A Marriage Made in Heaven

"No other primitive people has such an extensive and accurate record of its myths, tales, and legends as the North American Indian."[78]
"In the creation story, fruitfulness and light had come into the world because Morning Star and his realm of light had conquered and mated with Evening Star in her realm of darkness."[79]
"In theory the Skidi Pawnee ceremonies all have as their object the performance either through drama or through ritual of the acts which were performed in the mythologic age. The ritual is a formal method of restating the acts of the supernatural beings in early times..."[80]

How and when the Americas were first settled is lost in the mists of prehistory and remains a matter of much controversy and speculation. Whether the earliest inhabitants trekked across the Bering land-bridge which once connected Siberia with western North America, or whether they came in successive waves by way of boat, will not concern us here. What is certain is that sometime after their arrival from afar, the various tribes quickly set about exploring and expanding into the furthest outreaches of North and South America. Some, like those who settled along the Northwest coast of Canada and North America, practiced a relatively sedentary lifestyle centered around fishing and farming. Others, like the Plains Indians, eventually adopted a more nomadic lifestyle, following the buffalo herds wherever they might lead them.

---

[78] S. Thompson, *Tales of the North American Indians* (Bloomington, 1966), p. xvi.
[79] G. Weltfish, *The Lost Universe* (New York, 1965), p. 106.
[80] James Dorsey as cited by R. Linton, "The Origin of the Skidi Pawnee Sacrifice to the Morning Star," *American Anthropologist* 28 (1928), p. 461.

Among the tribes that Lewis and Clark encountered during their famous journey across the heartland of North America were the Skidi Pawnee, who had settled along the Loup river in what is now central Nebraska. The Skidi made their living hunting buffalo, raising corn, and raiding their neighbors.[81]

The Skidi comprise one of the four major bands of the Pawnee and are thought to have immigrated to the Midwestern plains from the South, perhaps preserving religious beliefs otherwise characteristic of the cultures of Mesoamerica and the American Southwest. They speak a Caddoan language.

At the time of their first encounter with Europeans—mostly Spanish and French trappers—the Skidi are thought to have numbered around 10,000. Within one century after the visit by Lewis and Clark, the population was reduced to some 600 individuals living on the brink of starvation and extinction.

The Skidi were inveterate sky-watchers. Indeed, it has been said that they were "obsessed with the planets"[82] and had "a sky oriented theology perhaps without parallel in human history."[83] The planet Venus was conceptualized as a Star Woman by the name of *cu-piritta-ka*, which translates literally as "female white star."[84] The anthropologist James Murie, himself of Skidi blood, summarized the lore surrounding this planet as follows:

"The second god Tirawahat placed in the heavens was Evening Star, known to the white people as Venus...She was a beautiful woman. By speaking and waving her hands she could perform wonders. Through this star and Morning Star all things were created. She is the mother of the Skiri. Through her it is possible for people to increase and crops to mature."[85]

It is to be noted that the planet Venus was explicitly distinguished from the "Morning Star." In fact, the Skidi identified the mythical "Morning Star" with the planet Mars, the latter envisaged as a powerful warrior of irascible disposition. Murie offered the following summary of the sacred traditions surrounding the Morning Star:

"The first one he placed in the heavens was Morning Star...This being was to stand on a hot bed of flint. He was to be dressed like a warrior and painted all over with red dust. His head was to be decked with soft down and he was to carry a war club. He was not a chief, but a

---

[81] For a general overview of their history, see B. Pritzker, *A Native American Encyclopedia* (Oxford, 2000), pp. 350-352.
[82] V. Del Chamberlain, *When Stars Came Down to Earth* (College Park, 1982), p. 82.
[83] *Ibid.*, p. 29.
[84] J. Murie, "Ceremonies of the Pawnee," *Smithsonian Contributions to Anthropology* 27 (Washington D. C., 1981), p. 39.
[85] *Ibid.*

warrior. ...Through him people were to be created and he would demand of the people an offering of a human being...He was also to be the great power on the east side of the Milky Way. This is Mars, u-pirikucu? (literally, 'big star'), or the god of war."[86]

Like other indigenous cultures, the Skidi traced their origins to the respective planets.[87] The central act of Skidi cosmogony described the Martian warrior's pursuit and eventual conquest of the planet Venus. Creation itself unfolded as a direct result of their sexual union. In summarizing the events in question, Ralph Linton stated simply "The Morning Star married the Evening Star."[88]

The sexual union between Mars and Venus featured prominently in numerous different aspects of Pawnee religious ritual. The *hieros gamos* between the planets was invoked every time a new fire was drilled, for example. In Skidi cosmology the drilling fire stick was identified with the prototypical masculine power (Mars as the "Morning Star") while the hearth symbolized the female power (Venus as the "Evening Star"). The firesticks themselves were addressed as follows: "You are to create new life even as Morning Star and Evening Star gave life to all things."[89] Murie summarized the symbolism of the fire-drilling ritual as follows:

"The Skiri also conceive of the firesticks as male and female. The idea is that the kindling of fire symbolized the vitalizing of the world as recounted in the creation. Specifically, the hearth represents the Evening Star and the drill the Morning Star in the act of creation."[90]

It will be noted that the Skidi skywatchers identified the hearth—the matrix of Creation—with the planet Venus. The prototypical fire-drill, on the other hand, was identified with the planet Mars. For the Skidi, as for indigenous cultures around the globe, the drilling of fire was conceptualized as a sexual act between cosmic powers.[91] Thus it is that, from a functional and symbolic

---

[86] *Ibid.*, p. 38.
[87] G. Dorsey & J. Murie, "Notes on Skidi Pawnee Society," *Field Museum of Natural History* 27:2 (1940), p. 77.
[88] R. Linton, "The Sacrifice to Morning Star by the Skidi Pawnee," *Leaflet Field Museum of Natural History, Department of Anthropology* 6 (1923), p. 5.
[89] J. Murie, *op. cit.*, p. 150.
[90] J. Murie, *op. cit.*, p. 40.
[91] R. Hall, *An Archaeology of the Soul* (Chicago, 1997), p. 98: "The drilling of fire by friction for the New Fire ceremony also symbolized the sexual union of Morning and Evening Star. The lower board or hearth board in such cases represented the Evening Star; the fire drill symbolized Morning Star." See also J. Frazer, *Myths of the Origin of Fire* (London, 1930), pp. 26ff.

standpoint, the ritual drilling of fire is analogous to a *hieros gamos* between Mars and Venus *in illo tempore*.

The *hieros gamos* between Mars and Venus also served as the mythological backdrop for human sacrifices. On rare occasions, or in the face of some perceived threat—the appearance of a meteor, an epidemic, or some other portent—the Pawnee offered a human sacrifice to the Morning Star, usually in the years when Mars appeared as a morning star.[92] Here a band of warriors would accompany a man impersonating the Morning Star in raiding a neighboring campsite, where they sought to kidnap a young woman of choice. Along the way there was much singing and dancing, during which the heroic deeds of the Martian warrior were recounted. Upon capturing a suitable victim, the war party returned to the Skidi village where several months might elapse while the priests prepared for the sacrifice and awaited signs for the most propitious time. The culmination of the rite saw the young woman—representing Venus—being painted head to toe and outfitted with a curious fan-shaped headdress. The victim was then led to a scaffold specially erected for the occasion whereupon, after mounting the final rung, she was shot in the heart by an arrow from the bow of the man impersonating Morning Star (see figure one).[93] The priests in charge of the gruesome rite took great care to ensure that the girl's blood was directed to a cavity below the scaffold. This pit was lined with white feathers and was held to represent the sacred garden of the planet-goddess: "The pit symbolized the Garden of the Evening Star from which all life originates."[94]

Figure 1

---

[92] R. Linton, "The Origin of the Skidi Pawnee Sacrifice to the Morning Star," *American Anthropologist* 28 (1928), p. 457. See also the detailed analysis by Von Del Chamberlain, *When Stars Came Down to Earth* (College Park, 1982).

[93] Adapted from E. Krupp, "Phases of Venus," *Griffith Observer* 56:12 (1992), with permission of the author.

[94] G. Weltfish, *The Lost Universe* (New York, 1965), p. 112.

In the Pawnee village, successful completion of the sacrifice was greeted with great rejoicing. According to Alice Fletcher's account, the men and women danced and "there was a period of ceremonial sexual license to promote fertility."[95]

As bizarre as this rite appears to the modern reader, anthropologists are generally agreed as to its fundamental purpose—to commemorate the sacred events of Creation and ensure fertility for the land. Ralph Linton's comments on the ritual are incisive in this regard:

"The sacrifice as a whole must be considered as a dramatization of the overcoming of the Evening Star by the Morning Star and their subsequent connection, from which sprang all life on earth. The girl upon the scaffold seems to have been conceived of as a personification or embodiment of the Evening Star surrounded by her powers. When she was overcome, the life of the earth was renewed, insuring universal fertility and increase."[96]

The Skidi traditions with respect to Venus and Mars raise a number of intriguing questions. How are we to explain the origin of such peculiar ideas and practices? The simplest explanation, as well as the most logical, is to trace the respective traditions to objective events involving the two planets. We would thus endorse the opinion expressed by the astronomer Ray Williamson: "The care with which the Pawnee observed the sky and noted the celestial events suggests that the story of Morning Star and Evening Star, in addition to serving as an explanation of the original events of the Pawnee universe, might also reflect actual celestial occurrences."[97]

The astronomer Von Del Chamberlain has conducted the most extensive investigation into the possible historical basis of the Skidi traditions.[98] He, too, concluded that astronomical events inspired the sacred traditions in question: "The conjunctions of Venus and Mars do seem to be the key to the Skidi concept of celestial parentage."[99] As for how these "conjunctions" are to be understood from an astronomical standpoint, Del Chamberlain surmised that they had reference to Mars' periodic migration from the morning sky to the western evening sky whereupon, on very rare occasions, it would conjoin with Venus. Other astronomers have since endorsed Del Chamberlain's interpretation.[100]

---

[95] *Ibid.*, p. 114.
[96] R. Linton, "The Sacrifice to Morning Star by the Skidi Pawnee," *Leaflet Field Museum of Natural History, Department of Anthropology* 6 (1923), p. 17.
[97] R. Williamson, *Living the Sky* (Norman, 1984), p. 225.
[98] V. Del Chamberlain, *When Stars Came Down to Earth* (College Park, 1982).
[99] *Ibid.*, p. 84.
[100] E. Krupp, *Beyond the Blue Horizon* (New York, 1991), pp. 189-192.

Granted that "actual celestial occurrences" are encoded in the Skidi myth of Creation, it remains far from obvious how we are to understand the origin of the specific motifs surrounding the two planets given Del Chamberlain's theory. Why was Venus conceptualized as the prototypical female power? Why was Mars viewed as masculine in nature or identified as Morning Star? Why would the periodic, relatively mundane, conjunction of these two particular planets be linked to Creation and ideas of universal fertility? Not one of these questions finds a satisfactory explanation under the thesis advanced by Del Chamberlain.

Perhaps the most important question facing students of ancient myth is the following: Do the Skidi myths with respect to Venus and Mars have an historical or observational basis? Stated another way: Are the sacred traditions in question to be understood as reliable memories of Creation and/or the recent history of the solar system, or are they a product of creative storytelling and thus unique to that particular culture?

It should be pointed out here that, at the time James Murie collected his anthropological reports from old medicine men and other eye-witnesses who had participated in the rites in question, the Skidi had been living in Indian Territory (Oklahoma) for over two decades, their previous way of life irreparably fractured and now only a distant memory. Douglas Parks characterized the degradation in Skidi culture as follows:

"The changes that occurred [with the move to Indian Territory and abandonment of traditional ways of life] were so basic that they had effectively destroyed the fabric of traditional Pawnee society. In no area of Pawnee life was this more poignantly true than in religion... Thus, when Murie began his work, he was observing a culture that was no longer viable and was only superficially approximating that of the surrounding white population. As time passed even more was irretrievably lost. By the first decade of the 20th century, when Murie's most systematic work was begun, the ceremonial knowledge of the remaining Skiri priests was largely fragmentary. The descriptions that he compiled are based on the memories of the old surviving priests, and are, as a consequence, uneven in quality. Other factors, such as the reluctance of some individuals to divulge their knowledge, undoubtedly contribute to this unevenness. In spite of the problems, however, Murie obtained an overview of the Skiri ceremonial calendar and filled out many parts of this scheme with an admirable amount of detail."[101]

---

[101] D. Parks, "Ceremonies of the Pawnee," in D. Parks ed., *Smithsonian Contributions to Anthropology* 27 (1981), p. 18.

In light of this history, it is to be expected that the traditions preserved by the Skidi informants might well have suffered disjuncture in certain respects. Yet as we will document, there is much reason to believe that the Skidi traditions with respect to Venus and Mars preserve an accurate memory of extraordinary events of earthshaking import. But how do we go about establishing this point?

In order to determine whether the Skidi astral traditions have an observational basis and thus preserve important information regarding the recent history of Venus and Mars, it is instructive to perform a cross-cultural analysis of astral lore. If the Skidi traditions have a rational foundation in actual historical events, they must find corroboration elsewhere. If, on the other hand, they are to be understood as fictional in nature or of relatively recent origin, it stands to reason that it would be most unlikely that cultures from the Old World would relate similar stories about the respective planets (that is, of course, unless they were directly influenced by Skidi beliefs). Yet if Old World cultures preserved myths and rites analogous to those from aboriginal North America, a *prima facie* case is thereby made for the thesis defended here, which holds that the Amerindian mythological traditions surrounding Venus and Mars encode and describe empirically observed catastrophic astronomical events.

The astronomical lore from ancient Mesopotamia offers a perfect case study in this regard inasmuch as it constitutes the earliest and most extensive body of traditions about the respective planets. Even from this brief summary it must be admitted that the Sumerian beliefs surrounding Inanna/Venus offer striking parallels to the Skidi traditions describing Venus. In addition to embodying the female principle, the planet is assigned a central role in a sacred *hieros gamos* thought to promote the fertility of the land.

The concordance between the Sumerian and Skidi traditions extends to the finest details. As the Skidi Venus was represented as a warrior-goddess threatening to throw the world into permanent darkness so, too, was Inanna/Venus invoked as an inveterate warmonger capable of turning light into darkness. Thus, an early hymn invokes Inanna as follows: "On the wide and silent plain, darkening the bright daylight, she turns midday into darkness."[102]

Despite the overwhelming evidence to the contrary, Sumerologists have generally sought to divorce Inanna's warrior-aspect from her identification with Venus.[103] Yet no matter how incongruous the image of a warring Venus might appear to the arm-chair theorist enthralled by the central dogma of modern

---

[102] Line 49 of "A Hymn to Inana," as quoted in J. Black et al, *The Literature of Ancient Sumer* (Oxford, 2004), p. 94.

[103] Gebhard Selz, "Five Divine Ladies," *NIN* 1 (2000), p. 34, writes: "It follows, that we have good reasons to doubt that the Sumerian Inana(k) was ever a proper war-goddess."

astronomy—i.e., that the planets have moved peacefully on their present orbits for countless million of years— Inanna's destructive demeanor is inherent in her archetypal manifestation as the planet Venus. This is stated explicitly in Iddin-Dagan's sacred marriage hymn, as elsewhere:

"As the lady, admired by the Land, the lone star, the Venus star, the lady elevated as high as the heaven, descends from above like a warrior, all the lands tremble before her."[104]

The life-giving garden associated with the Skidi planet-goddess also finds a symbolic counterpart in Sumerian tradition. Thus a garden of Inanna/Venus is mentioned in conjunction with the sacred marriage rite:

"A garden of the goddess ($kiri_6$-nin-ku_3-nun-na) is attested in the oldest extant ritual text…According to lines 9-11 of this ritual the king is to bathe in the garden on the night of the fourth day of the ritual."[105]

As the Skidi held that "all life" originated from Venus's sacred garden so, too, did the Sumerians deem the planet Venus to be the "divine source of all life."[106] This is but one of hundreds of archetypal motifs associated with Venus that will never be explained by reference to the distant planet familiar to modern astronomers.

The most comprehensive study of the sacred marriage rite in ancient Mesopotamia is that by Pirjo Lapinkivi. She poses the following question:

"The language of most of the sacred marriage texts is so explicitly sexual that it seems beyond question that they describe a sexual union between the king and the goddess Inanna, the consummation of their marriage. The crucial question, however, is, why? Why did this union take place, and why was it performed ritually…?"[107]

Lapinkivi then proceeds to answer her own question—the historical origin and fundamental purpose of the sacred marriage rite remains unknown:

"Despite all the various suggestions reviewed above, no scholarly consensus has been reached regarding this basic question. While the importance of the sacred marriage for the Sumerians is obvious, it has remained enigmatic to the modern scholars."[108]

---

[104] Lines 135ff. as quoted in J. Black et al., *op. cit.*, p. 266.

[105] M. Hall, *A Study of the Sumerian Moon-God, Nanna/Suen,* PhD dissertation University of Pennsylvania, 1985, pp. 750-751.

[106] F. Bruschweiler, *Inanna. La déesse triomphante et vaincue dans la cosmologie sumérienne* (Leuven, 1988), p. 112. See also the discussion in B. Hruška, "Das spätbabylonische Lehrgedict 'Inanna's Erhöhung'," *Archiv Orientalni* 37 (1969), p. 482: "In der sumerischen Zeile wird ᵈištar-kakkabi mit dem Namen ti-mú-a 'Leben erzeugende' wiedergegeben."

[107] P. Lapinkivi, *The Sumerian Sacred Marriage* (Helsinki, 2004), p. 14.

[108] *Ibid.*, p. 14.

It doesn't take a rocket scientist to understand why scholars have failed to discern the original significance of the sacred marriage rite: They have all but ignored the formative influence of extraordinary celestial events in the genesis of ancient myth and religion. Thus it is that the all-important role of the planet Venus in the sacred marriage rite has been essentially overlooked. The fact that most scholars have eschewed a comparative approach has also proven myopic and prevented them from discovering that analogous traditions surround Venus in other cultures.

The sacred lore surrounding Inanna prompts a host of questions. Why would the early kings of Mesopotamia seek legitimization for their rule through a symbolic marriage with the planet Venus, the latter personified by Inanna? How are we to understand the curious belief maintaining that the king's sexual union with Inanna/Venus would promote fertility and abundance throughout the land? Why is the planet Venus involved in a "sacred marriage" at all, much less one with a supposedly mortal king like Dumuzi? And if Inanna is to be identified with the planet Venus, how are we to understand her mortal paramour Dumuzi? Was he, too, originally a celestial body and, if so, which one? Although such questions cry out for explanation, it must be said that few scholars have sought to address them in any sort of systematic manner.

4. The Many Loves of Aphrodite

"Blessed Queen of Heaven…celestial Venus, now adored at sea-girt Paphos, who at the time of the first Creation coupled the sexes in mutual love."[109]

"Ares and Aphrodite serve as the mythical paradigm for marriage."[110]

In ancient Greece, as today, the name Aphrodite evoked images of alluring beauty and erotic passion. Yet an aura of mystery surrounds the Greek goddess of love. Walter Burkert, upon surveying the evidence, confesses: "Aphrodite's origin remains as obscure as her name."[111] While Aphrodite is securely attested in the earliest epic literature, her name is absent from the Mycenaean religion as known from the Linear B tablets. Thus, it is likely that the cult of the goddess came to Greece at some point during the period between 1200 BCE and 800.[112]

Whence, then, did Aphrodite arrive on Greek shores? For Homer, Hesiod, and other early writers, the goddess was intimately linked to Cyprus. The *Odyssey* lists Paphos as the goddess's homeland, while the *Iliad* makes *Kypris* her most common epithet.[113] Hesiod calls her both *Kyprogene* and *Kythereia*.

The search for Aphrodite's origins does not stop in Cyprus, a well-known melting pot of Oriental religious conceptions. Among leading scholars, there is

---

[109] Apuleius, *The Golden Ass* (New York, 2009), p. 262
[110] L. Reitzammer, *The Athenian Adonia in Context* (Madison, 2016), p. 176.
[111] W. Burkert, *Greek Religion* (Cambridge, 1985), p. 153.
[112] C. Penglase, *Greek Myths and Mesopotamia* (London, 1994), pp. 176ff.
[113] According to C. Penglase, *op. cit.*, p. 176, "The earliest evidence for Aphrodite in the Greek and Mycenaean area is the temple in Paphos." See also W. Burkert, *op. cit.*, p. 153.

something of a consensus that the cult of Aphrodite originally came to Greece from the ancient Near East: "Behind the figure of Aphrodite there clearly stands the ancient Semitic goddess of love, Ishtar-Astarte, divine consort of the king, queen of heaven, and hetaera in one."[114]

This view receives strong support from the Greeks themselves. Pausanias, for example, offered the following opinion: "The Assyrians were the first of the human race to worship the heavenly one [Aphrodite *Urania*]; then the people of Paphos in Cyprus, and of Phoenician Askalon in Palestine, and the people of Kythera, who learnt her worship from the Phoenicians."[115]

That Aphrodite shares numerous characteristics in common with Ishtar is well known. As goddesses of love both are associated with rites of prostitution, for example.[116] Both are associated with sacred gardens. Aphrodite, like Ishtar, was represented as armed and invoked to guarantee victory. The strange beard accorded Aphrodite in ancient cult finds a precise parallel in the cult of Ishtar.[117]

In his comprehensive survey of Aphrodite's cult, Burkert never once mentions the planet Venus. Here the renowned scholar is presumably just following the currently prevailing view, which does not recognize an early connection between the goddess and planet (the identification between Aphrodite and Venus is first attested in the *Epinomis*, now generally ascribed to Philip of Opus). Wolfgang Heimpel's opinion on this issue seems to represent the consensus among Classicists:

"Originally, the goddess Aphrodite had nothing to do with the planet. The link was in all probability made as a result of Babylonian influence in the field of astronomy."[118]

Yet inasmuch as the Semitic Ishtar was specifically identified with Venus, it stands to reason that the Greek goddess originally shared this identification as well. In fact, it is our opinion that it is impossible to understand Aphrodite's cult and mythology apart from reference to the planet Venus.[119]

In order to determine whether Aphrodite's identification with Venus is archaic in nature or relatively late in origin—as per the view of Heimpel and the vast majority of scholars—it is necessary to investigate her cult in some detail.

---

[114] W. Burkert, *op. cit.*, p. 152.
[115] Book I:14:7.
[116] C. Penglase, *op. cit.*, p. 163, citing Strabo 378 for Corinthian cults of prostitution associated with Aphrodite. Notice also the epithet *Porne*.
[117] M. Jastrow, "The Bearded Venus," *Revue Archéologique* 17 (1911), pp. 271-298.
[118] W. Heimpel, "A Catalog of Near Eastern Venus Deities," *Syro-Mesopotamian Studies* 4:3 (1982), p. 11.
[119] E. Cochrane, *Starf*cker* (Ames, 2006), pp. 43-72.

Aphrodite's epithet *Urania* offers a valuable clue. As Farnell points out,[120] *Urania*—"the celestial one"—was a Greek translation of the Hebrew epithet *malkat ha-šāmayim*, "the queen of the heavens," long understood as having reference to Venus.[121] Yet Farnell still questions whether Aphrodite's epithet betrays an astral component! Such an opinion ignores the plain fact that this epithet finds precise parallels in the cults of other Venus-goddesses throughout the ancient world. Thus, a Sumerian hymn invokes Inanna as follows:

"*I shall greet the great lady of heaven, Inana! I shall greet the holy torch who fills the heavens, the light, Inana, her who shines like daylight...the respected one who fills heaven and earth with her huge brilliance.*"[122]

The Akkadian Ishtar shares the same epithet. Witness the following hymn:

"*Her very first name, her great appellation which her father Anu, whom she adores, named her of old, is Ninanna 'Queen of Heaven'.*"[123]

How is it possible to understand these early hymns to Inanna and Ishtar apart from reference to a celestial body?

The Queen of Heaven also figures prominently amongst the pagan gods mentioned in the Old Testament, and there was doubtless much truth in Jeremiah's admission (ca. 600 BCE) that the Israelites had long burnt incense to the stellar whore.[124] Although Jeremiah does not name the planetary goddess in question, Astarte seems a likely candidate.[125] Astarte's identification with the planet Venus is commonly acknowledged, as is her fundamental affinity with Aphrodite.[126] Indeed, an inscription from the Hellenistic period (ca. 160 BCE) identifies Astarte with Aphrodite *Urania*.[127] Given this evidence from comparative religion, there would appear to be little justification for Farnell's view that Aphrodite *Urania* did not have a celestial component.

In Greek myth Aphrodite is known primarily for her liaisons with various gods and heroes. A famous passage in Homer's *Odyssey* finds the "widely renowned" myth-maker Demodocus singing of the illicit love affair of Ares (Mars) and Aphrodite (Venus). In the song in question, the two lovers are entrapped *in*

---

[120] L. Farnell, *The Cults of the Greek States, Vol. II* (New Rochelle, 1977), p. 629.

[121] See L. Bobrova & A. Militarev, "From Mesopotamia to Greece: to the Origin of Semitic and Greek Star Names," ed. by H. Galter, *Die Rolle der Astronomie in den Kulturen Mesopotamiens* (Graz, 1993), p. 315.

[122] Lines 1-8 from "A šir-namursaga to Ninsiana for Iddin-Dagan (Iddin-Dagan A)," *ETCSL*.

[123] B. Foster, *Before the Muses* (Bethesda, 1993), p. 501.

[124] *Jeremiah* 44:17-25.

[125] W. Heimpel, *op. cit.*, p. 21.

[126] J. Henninger, "Zum Problem der Venussterngottheit bei den Semiten," *Anthropos* 71 (1976), pp. 153ff. See also M. Astour, *Hellenosemitica* (Leiden, 1967), p. 116.

[127] W. Heimpel, *op. cit.*, p. 21.

flagrante delecto in an invisible net devised by the lame smith Hephaistos, much to the amusement of the other gods, who are witness to the entire scene:

> *"Then playing the lyre, the man struck up a beautiful song of the love of Ares and fine-crowned Aphrodite, how they first had lain together in the home of Hephaistos, secretly. He gave her many gifts and shamed the marriage bed of her lord Hephaistos…The two of them went to sleep in the bed clothes. And the bonds fashioned by various-minded Hephaistos were spread about them, so they could not move or raise their limbs at all. And then they knew it when they could no longer escape."*

Upon ensnaring the two lovers, Hephaistos—the goddess's husband—bemoans his fate:

> *"Father Zeus and you other blessed ever-living gods, come see deeds to laugh at that are not be endured, how Aphrodite, the daughter of Zeus, dishonors me always, lame as I am, and loves the destructive Ares, because he is handsome and nimble of foot, while I was born feeble."*[128]

For well over two thousand years now, scholars and laymen alike have been scratching their heads trying to figure out whence Homer derived this bawdy tale, one which the ancients themselves found more than a little disconcerting if not downright blasphemous.[129] Was the love affair between Aphrodite and Ares a product of the blind bard's vivid imagination, or a traditional tale deeply rooted in ancient religion and ritual?

The available evidence suggests that the sexual union between Ares and Aphrodite was of hoary antiquity. Certainly the two gods were thoroughly enmeshed in Greek cult and ritual as well as in Homer.[130]

The testimony of the early Greek poets and dramatists supports this conclusion. Aeschylus describes Ares as the "Manslayer, who beds

---

[128] *Odyssey* VIII:266-311, as translated in A. Cook, *The Odyssey* (New York, 1974), p. 106.

[129] Witness the commentary of Anne Teffeteller, "The Song of Ares and Aphrodite," in A. Smith & S. Pickup eds., *Brill's Companion to Aphrodite* (Leiden, 2010), p. 137: "Demodokos' song of Ares and Aphrodite in Homer's *Odyssey*, a narrative which delighted Odysseus and the Phaiakians but mortified the ancient moralists and has perplexed many readers of Homer down to the present day." In his commentary on the Odyssey, W. B. Stanford, *The Odyssey of Homer, Vol. 1* (London, 1947), p. 338 observes: "The following ballad of Ares and Aphrodite has been much criticized by both higher critics and moralists."

[130] W. Burkert, "The Song of Ares and Aphrodite," in L. Doherty ed., *Homer's Odyssey* (Oxford, 2009), p. 29 observes: "Admittedly, the connection between Ares and Aphrodite is firmly rooted in cult and myth." So, too, Jennifer Larson, *Ancient Greek Cults* (New York, 2016), p. 116 notes that Aphrodite's relationship with Ares "appears to be an archaic feature of her worship rather than a late development."

Aphrodite."[131] Pindar refers to Ares as Aphrodite's husband: "Ares, the husband of Aphrodite."[132] The same poet elsewhere described Aphrodite as "mother of loves in the sky"—suggestive evidence that the Theban poet was fully cognizant of the fact that the celebrated love affair of Aphrodite and Ares had its origin *in the sky*.[133]

For Euripides, Aphrodite was a principal catalyst in Creation. This idea is stated most clearly in *Hippolytus*: "Everything is generated by her, she is the one who sows and gives desire, from which all of us on earth exist."[134] This sentiment is echoed in Apuleius: "Blessed Queen of Heaven...celestial Venus, now adored at sea-girt Paphos, who at the time of the first Creation coupled the sexes in mutual love."[135] Significantly, the very same idea is also evident in Skidi cosmogonic myth. Recall the passage quoted earlier: "Through this star [Venus] and Morning Star [Mars] all things were created."[136] It is at this point that the critically-minded reader must ask themselves: What are the odds that two disparate cultures on the opposite side of the Atlantic Ocean would conceptualize a distant planet as the Creatrix of all living things?

Aphrodite is the marriage-goddess *par excellence*. In his play *Phaethon*, preserved today only as a collection of fragments, Euripides describes Aphrodite as follows: "We sing the heavenly daughter of Zeus, the mother of loves, Aphrodite, who brings nuptials to maidens."[137] Yet as James Diggle notes, there is also an explicit connection to the planet Venus: "[Venus is] the star which in poetry enjoys a close association with Aphrodite and with the marriage ceremonial."[138]

Our earliest and likely most reliable witness regarding Greek matrimonial practices is the melancholy poet Sappho, who wrote around 600 BCE from her island homeland Lesbos.[139] For Sappho, it was Aphrodite herself who

---

[131] *Suppliants* 659-660.
[132] *Pythian Odes* 4.87ff.
[133] Fragment 122:3-5. See discussion in C. Collard et al eds., *Euripides: Selected Fragmentary Plays, Vol. 1* (Oxford, 1995), p. 235.
[134] *Hippolytus* 448-450.
[135] Apuleius, *The Golden Ass* (New York, 2009), p. 262
[136] *Ibid*.
[137] Lines 229-230.
[138] J. Diggle, *Euripides: Phaethon* (Cambridge, 1970), p. 12.
[139] According to L. Reitzammer, *The Athenian Adonia in Context* (Madison, 2016), p. 46 "Although Sappho's poetry is removed in space and time from Classical Athens, her *epithalamia* [wedding songs] fragments are our best-preserved examples of songs sung at weddings. In general, the images and metaphors associated with Greek weddings remain relatively consistent over time."

represented the archetypal bride.[140] The raging war-god Ares, in turn, she describes as the divine prototype of bridegrooms. Thus a recently discovered fragment reads as follows:

"A bridegroom will come equal to Ares, hymenaios! Bigger by far than a big man (fr. 111 Voigt)."[141]

A singular detail in Sappho's fragmentary account finds the bridegroom being infused with the divine charisma of the celestial goddess. Gregory Nagy summarized the available evidence as follows:

"In the wedding songs of Sappho, the god Ares is a model for the generic gambros, 'bridegroom', who is explicitly described as isos Areui, 'equal [isos] to Ares,' in Sappho Song 111.5. Correspondingly, there are many instances of implicit equations of the generic bride with the goddess Aphrodite: in Sappho Song 112, for example, the bridegroom is said to be infused with the divine charisma of Aphrodite evidently by way of his direct contact with the bride."[142]

This line of inquiry prompts a number of questions. The most obvious, perhaps, is the following: To what extent were such traditions inspired by the witnessed behavior of the planets Mars and Venus? To even ponder such a possibility is to risk ridicule in classical circles—not to mention in astronomical departments around the globe.

As it turns out, Sappho's peculiar report regarding Aphrodite's infusion of her Ares-like bridegroom with charisma offers a decisive clue inasmuch as analogous traditions will be found around the globe, typically in connection with the planet Venus. If this claim can be substantiated, we will have gone a long way towards establishing the fact that Sappho was likely transmitting ancient traditions of planetary interactions when she spoke of the archetypal roles of Aphrodite and Ares.

The sacred lore from ancient Persia is especially instructive here: According to the *Aban Yašt* from the *Zend-Avesta*, the planet Venus was identified with the goddess Anahita.[143] Anahita, much like the Greek Aphrodite *Urania*, was intimately associated with ancient conceptions of sovereignty, kingship, and fertility. Indeed, the planet-goddess's connection with kingship was so close that Persian coronation scenes depicted her handing the king his crown.[144] Most

---

[140] See the discussion in G. Nagy, *The Greek Hero* (Cambridge, 2013), p. 118.
[141] L. Reitzammer, *op. cit.*, p. 41.
[142] G. Nagy, *op. cit.*, p. 118.
[143] *Yašt* 5:85. See also *Bundahishn* 5:1.
[144] Y. Ustinova, "Aphrodite Urania," *Kernos* 11 (1998), p. 218. See also B. Marshak, "Pre-Islamic Painting of the Iranian Peoples and its Sources in Sculpture and the Decorative Arts," in E. Sims ed., *Peerless Images: Persian Painting and Its Sources* (New Haven, 2002),

significant, however, is the fact that the goddess was believed to invest the king with *charisma*: "She legitimated the enthronement of the king, providing him with charisma."[145]

As we have documented elsewhere, such ideas were surprisingly widespread throughout the ancient world.[146] Among various cultures of Inner Asia, for example, the charisma bequeathed by the planet-goddess was a central theme:

"To become a real ruler, the Hunnic Chanyu had to possess sacral grace. Through it, he ensured the welfare of his people, as well as fertility, successful military campaigns, etc. This supernatural power was passed on by inheritance, and only in the ruling family. Among the Iranians, the notion of sacral grace (charisma) is expressed with the concept of hvarna, farn, and among the Turks—with the concept of qut. In ancient Iran and Turan, the rulers fought for this charisma and received it from Aredvi Sura Anahita. Its possession is also related to the Turanian sacred mountain in Kanha, which was probably a religious center and part of the Anahita cult. A similar belief has been preserved among the Turks: qut could also be obtained from the sacred mountain Otuken, regarded as a female deity. Thus, the Iranians and Turks both believed that the most important quality a ruler needed to have to be able to govern was the result of the blessing and support of the Great Goddess."[147]

To reiterate: In order to assume the throne and ensure general welfare and fertility, the king must first receive the "blessing" and divine charisma of the Queen of Heaven, Venus. This is getting curiouser and curiouser, as they say. In what sense is it possible to understand the planet Venus as infusing the ancient king or bridegroom with charisma? What, exactly, is charisma?

The Greek word *charisma* derives from *charis*, a word often translated simply as "grace" or splendor but attested in a wide range of different, albeit related, meanings. Thus *charis* denotes not only the beauty of Aphrodite or any other young woman, it can also describe the "glory" or undying fame awarded to the winning athlete after some great contest.[148] For the ancient Greeks, the *charis* was thought to depart at death.[149]

It is possible to be more specific here: *charis* describes an effulgent aura or nimbus-like crown encircling another object. Indeed, according to an insightful

---

p. 10.
[145] E. Yarshater, *The Cambridge History of Iran, Vol. 2* (Cambridge, 1980), p. 846.
[146] E. Cochrane, *Starf*cker* (Ames, 2006), pp. 152-158.
[147] B. Zhivkov, *Khazaria in the Ninth and Tenth Centuries* (Leiden, 2015), p. 101.
[148] B. MacLachlan, *The Age of Grace: Charis in Early Greek Poetry* (Princeton, 1993), p. 5.
[149] *Ibid.*, p. 4: "At death one faced the dreary prospect of the disappearance of *charis*."

analysis by Richard Onians, Homer conceptualized Charis "as a wreath or crown about things."[150]

The Avestan word *hvarna*—also spelled *Xvarenah*, *khvarna*, and *farn/ farr*—is translated alternately as "the charisma of Fortune" or "sovereign glory."[151] The radiant phenomenon in question is understood to be a supernatural power substance with which the king is invested but which might also be taken away from him, resulting in disaster for the king and threatening the world with destruction.[152] In Persian myth, the primordial king Yima is represented as presiding over a veritable Golden Age of riches and prosperity until he began to dabble in falsehood and deceit, at which point his luminous Glory departed from him, thereby plunging the world into chaos. Thraetona, the greatest Persian hero of them all, eventually succeeded in regaining the *hvarenah* and, upon enveloping himself with it, proceeded to save the world from imminent destruction.[153]

Persian kings, in a purposeful attempt to emulate Thraetona's heroics, sought to provide themselves with the selfsame "glory" in order to gain the throne and rule in power: "In assuming the throne, however, he [the Great King] took unto himself the mystique, spirit and glory of kingship: tradition has it that he was crowned on his birthday, at which event he was thought to be reborn, and thus, assumed a throne name."[154] Bendt Alster is more specific: "Where tradition is still more or less a living thing, great monarchs consider themselves imitators of the primordial hero: Darius saw himself as a new Thraetona."[155]

Additional insight into the luminous charisma associated with the planet Venus can be obtained from the earliest religious traditions of the Near East. For the ancient cultures of Mesopotamia the union of Inanna and Dumuzi symbolized the divine paradigm of marriage, much as the union of Aphrodite and Ares did for the Greeks. In dozens of ancient hymns, Inanna/Venus is represented as the

---

[150] R. Onians, *The Origins of European Thought* (Cambridge, 1951), p. 402.

[151] M. West, *Indo-European Poetry and Myth* (Oxford, 2007), p. 271 translates the term simply as "sovereign glory."

[152] See the discussion in J. Gonda, *Some Observations on the Relations Between 'Gods' and 'Powers' in the Veda* ('S-Gravenhage, 1957), pp. 14-23. See also J. Puhvel, *Comparative Mythology* (Baltimore, 1987), p. 106.

[153] A. Carnoy, "Iranian Views of Origins in Connection with Similar Babylonian Beliefs," *Journal of American Oriental Studies* 36 (1917), p. 308.

[154] T. Young, "The consolidation of empire and its limits of growth under Darius and Xerxes," in J. Boardman et al eds., *The Cambridge Ancient History IV* (Cambridge, 1988), p. 105.

[155] B. Alster, "The Paradigmatic Character of Mesopotamian Heroes," *Revue d'assyriologie et d'archeologie orientale* 68 (1974), p. 51.

exemplary "bride" and Dumuzi as the prototypical bridegroom.[156] Famously, an archaic Sumerian ritual found the king impersonating Dumuzi and simulating a sexual union with the planet Venus (as Inanna) in order to legitimate his hold on the throne and secure fertility throughout the land.[157] According to various early literary accounts of the rite in question, a key episode found the planet-goddess imbuing her royal bridegroom with luminous splendor or power. This idea is evident in the following passage from the Old Babylonian hymn BM 96739:

"Oh Inanna, a husband worthy of your splendor has been granted to you...You, oh mistress, you have handed over to him your power as is due to a king, and Ama-ušumgal-anna causes a radiant brilliance to burst out for you."[158]

Françoise Bruschweiler, in her masterful analysis of the symbolism associated with Inanna/Venus, offered the following commentary on this particular hymn:

"This passage is interesting due to the way in which, in the context of a sacred marriage, the luminous essence of the goddess is passed over to King Ama-ušum-gal-anna, who is identified for the occasion with Dumuzi."[159]

In an early hymn Inanna/Venus is invoked as ni_2-gal, a phrase commonly translated as "Great awesomeness," "glory," or "splendor" by leading Sumerologists.[160] That the glory in question is celestial in nature is confirmed by the following passage: "Inana, you diffuse awesomeness like fire" [throughout heaven]."[161] Elsewhere Inanna boasts that she has imbued the king with this extraterrestrial "glory" (translated here by the innocuous word "awesomeness"):

"Imbued (?) with my awesomeness! Imbued (?) with my awesomeness! The life of the lord, imbued with my awesomeness! The life of the king, imbued (?) with my awesomeness!"[162]

---

[156] Lines B: 16-17 from "A song of Inana and Dumuzid (Dumuzid-Inana Z), *ETCSL*: "Let us embrace, my bridegroom! Let us lie on my flowered bed!"

[157] Y. Sefati, *Love Songs in Sumerian Literature* (Jerusalem, 1998), p. 49: "The king's union with the goddess resulted in her granting a favorable promise of fertility and abundance for the land and its inhabitants."

[158] Quoted from F. Bruschweiler, *Inanna. La déesse triomphante et vaincue dans la cosmologie sumérienne* (Leuven, 1988), p. 110. For a slightly different translation, see J. Black et al., "A tigi to Inana (Inana E)," *ETCSL*, lines 21-24: "Mistress [Inanna], you have given your strength to him who is king. Ama-ušumgal-ana brings forth radiance for you."

[159] *Ibid.*, p. 111.

[160] See line 186 from "A hymn to Inana (Inana C)," *ETCSL*.

[161] Line 120 from "A hymn to Inana as Ninegala (Inana D)," *ETCSL*.

[162] Lines B:10-12 from "A šir-namšub to Inana (Inana I)," *ETCSL*.

The natural history reflected in this archaic tradition is as simple as it is decisive for the proper interpretation of ancient conceptions of kingship and the curious belief-systems attached to marriage rituals: The royal "glory" or *charisma* is nothing other than the radiance of the planet Venus, universally identified as the prototypical female celestial power. So long as the prototypical king had his Venusian aura, all is well and fertility abounds throughout the land. But as soon as the "glory" departs from the universal sovereign disaster overtakes the world and chaos and darkness reign.

With regards to the specific details of the planetary conjunction encoded in ancient traditions of Aphrodite imbuing Ares with "charisma" or Inanna investing the king (as Dumuzi/Mars) with ni_2-gal, it is important to understand that our model requires that the red planet be positioned squarely *in front of* the larger Venus within the so-called polar configuration of planets (see the illustration in figure 1). It was this decidedly extraordinary conjunction of planetary powers that was conceptualized as the "marriage" of Venus and Mars or, alternately, as Venus's imbuing of the red planet with charisma in the guise of the luminous crown of kingship.[163] As a result of these spectacular natural events, the red planet was viewed as having gained "Sovereignty" as King of the Gods. Indeed, it is precisely because the marriage of Venus and Mars is functionally and structurally analogous to the "crowning" of Mars that we would understand the indissoluble connection between the sacred marriage rite and kingship. Thus it is that, from the standpoint of historical origins, to be "King" meant nothing less than to be "married" or conjoined with the planet Venus. For it was solely by means of his union with the planet Venus that the King of the Gods acquired the charismatic "glory" and ascended to the throne.

It goes without saying that the unique conjunction of planets depicted in figure one is quite impossible in the current solar system due to the fact that Mars, as an outer planet, can never appear "in front" of Venus, an interior planet. Hence the bold challenge to conventional ideas of astronomy presented by our radical historical reconstruction based primarily on the written testimony of the ancient skywatchers themselves.

Figure one

---

[163] D. Talbott, *Symbols of An Alien Sky* (Portland, 1997), pp. 92-102.

5. Aphrodite and Phaon

"In fact, the traditions of Greek lyric are in many ways older than the traditions of Greek epic, and the myths conveyed by epic are in many ways newer than the myths conveyed by lyric."[164]

If the cult of Aphrodite encodes ancient conceptions associated with the planet Venus, it must be expected that astronomical phenomena will inform and help illuminate specific details in the sacred traditions surrounding the goddess. In order to investigate this hypothesis we offer a comparative analysis of Aphrodite's rendezvous with Phaon.

A curious story, popular in Greek comedy and preserved in fragmentary fashion by various ancient writers, relates that Aphrodite once befriended an old ferryman named Phaon after the latter had aided the goddess in crossing the Aegean. In return for his random act of kindness, the goddess rewarded the old man by magically transforming him into a handsome youth.[165]

In addition to these basic facts, there are also hints that Aphrodite and Phaon were lovers. Thus, Athenaios reports that the Cytherean goddess was in love with the ferryman, citing Kratinos, Euboulos and Kallimachos as authorities.[166] Kratinos wrote that Phaon was the most beautiful man on earth and that Aphrodite had hidden him away in order to keep him for herself.[167]

---

[164] G. Nagy, "Homer and Greek Myth," in R. Woodard ed., *The Cambridge Companion to Greek Mythology* (Cambridge, 2007), p. 52.

[165] Sappho fragment 211 V. G. Nagy, "Phaethon, Sappho's Phaon, and the White Rock of Leukas," *Harvard Studies in Classical Philology* 77 (1973), p. 177, writes simply that Aphrodite conferred "youth and beauty on Phaon."

[166] Athenaios, *Deipnosophistae* 2.69d.

[167] PCG IV fragment 370; Kallimachus fragment 478. See also L. Köppel, "Phaon," in H.

Although Palaephatos (late 4th century BCE) is our earliest source for the story in question, Menander (ca. 324 BCE) and other writers also allude to it.[168] Aphrodite's encounter with Phaon is also depicted on several vase paintings.[169]

Considered in isolation, it is difficult to make much headway in deciphering the original significance of these fragmentary traditions from ancient Greece. Certainly it is far from obvious that planetary interactions hold the key to Aphrodite's tryst with Phaon. For additional insight into the origins of the Greek legend we turn to consider sacred traditions from aboriginal South America.

A fascinating myth, widespread in South America, is the so-called "Star Woman" cycle (A762.2 in Thompson's Index). The basic plot finds a beautiful star visiting Earth and carrying off a mortal to make her lover or husband. In most versions of the tale, the mortal paramour is distinguished by his old age, ugliness, or some deformity, yet as a result of his union with the Star Woman he is magically transformed into a handsome youth. Occasionally it is reported that the Star Woman and her lover ascend to heaven and live happily ever after. A few examples of this myth will serve to illustrate its relevance for understanding the Greek tale of Aphrodite and Phaon.

In the first decade of the 20th century, Alberto Fric became the first white man to record a sampling of Chamacoco lore (the latter tribe hails from the Paraguayan Chaco). Included in his collection is the following narrative telling of a Star Woman's love for a homely mortal:

"Formerly the star Venus was a woman who fell in love with a homely man. Thanks to her magic, he became very handsome."[170]

Several different versions of this story were subsequently obtained from other Chamacoco informants. Although most are more elaborate and embellished than Fric's brief account, the same basic plot is usually recognizable. In their compendium of Chamacoco lore, Wilbert and Simoneau include a version narrated by Bruno Barras, the highlights of which are as follows:

"Once there was a bachelor. Every night when he lay down to sleep he wished he had a beautiful wife, a fair-skinned wife. Lying in bed at night he would see the star called Iozly [Venus]...Then the star came...She

---

Cancik & H. Schneider eds., *Der Neue Pauly* 9 (Stuttgart, 2000), col. 736.

[168] Palaephatos 48. The best summary of the extant sources is that of Stein, "Phaon," *RE* 38 (Stuttgart, 1938), cols. 1790-1796. See also T. Gantz, *Early Greek Myth* (Baltimore, 1993), pp. 103-104.

[169] *Lexicon Iconographicum Mythologiae Classicae, Vol. 7* (Zurich, 1994), pp. 364-367.

[170] J. Wilbert & K. Simoneau, *Folk Literature of the Chamacoco Indians* (Los Angeles, 1987), p. 97.

said: 'Don't be afraid. Because you have been looking at me year after year I have now come to sleep with you. I want to be your wife…'[171]

In the ensuing weeks, Star Woman continued to make nocturnal visits to Earth. The natives eventually grew restless and more than a little jealous at the dramatic transformation in the bachelor's appearance:

"When she lay down with him there was a light emanating from her, illuminating everything…By now the other people and some girls were very envious of the young man's family because they looked so fair and beautiful. The man used to be dark and ugly, but when he slept with Iozly every night he grew better and better looking until he was fair and handsome, with smooth, fair hair."[172]

The Star Woman cycle is widely distributed amongst the various tribes native to the Gran Chaco region, including the Apinaye, Chorote, Makka, Mocov', and Toba. Of the Star Woman myth in general, the anthropologist Alfred Métraux wrote: "This tale is very popular with Chaco Indians, and it is generally the first story they tell when asked about their folklore."[173]

A Chorote version of the myth serves to complement the Chamococo narrative. Here, too, a mortal of homely appearance formed the object of Venus's affections:

"There was a man who was so ugly that no woman wanted him. All the women in his village persecuted him, throwing sticks at him. At night he lay down to sleep outside and started to look up at Katés: 'What a pretty girl! How I should like to marry her!'…The following night Katés descended to the earth and had intercourse with the young man. When dawn was near she said to him: 'I come from the sky, and at night I shall be your wife. Do not tell anybody that I have come. I do not go about during the daytime, and so that no one will see me I am going to hide inside that gourd.'"[174]

Another Chorote informant offered a slightly different version of the Star Woman narrative. It begins as follows:

"In primordial times, a young man was outside every night, looking at beautiful stars, for the stars were women. He especially looked at Katés (Morning Star), thinking: 'I should like her to be my wife.'"[175]

---

[171] *Ibid.*, p. 85.
[172] *Ibid.*, pp. 85-86.
[173] A. Métraux, *Myths of the Toba and Pilagá Indians of the Gran Chaco* (Philadelphia, 1946), p. 46.
[174] J. Wilbert & K. Simoneau, *Folk Literature of the Chorote Indians* (Los Angeles, 1985), pp. 265-266.
[175] *Ibid.*, p. 257.

The youth had first gained Star Woman's attention by shooting an arrow at her. As a result of this affront she promises to visit him:

"Exactly at midnight the woman came. Now he had a wife. In the morning everyone looked at the young man whom nobody had wanted previously. No girl from his village liked him."[176]

In order to keep their affair a secret from the other tribesmen, Star Woman asks her mortal lover to find a gourd so she can enter into it and remain concealed from sight. Eventually, following further adventures—one of which found Star Woman forced to reconstitute her dismembered lover's body—"she took him with her to the sky where she lives."[177]

The Ge of Central Brazil tell a very similar story. As recorded by Wilbert and Simoneau, the narrative begins as follows:

"A boy was lying down in the middle of the plaza, and Katxere was looking down at him. She felt sorry, and said: 'I am going to marry that boy.'"[178]

After sleeping with the boy, Star Woman tells him to hide her in a basket (*kaipo*) or gourd (*combuca*).[179]

A Toba narrative preserves the same basic story but adds a few interesting plot-twists. Explicitly identified with the planet Venus, Star Woman is described as having "long hair."[180] As in other versions of the tale, the mortal hero "hid her in a large gourd so that no one would see her."[181] Once again Star Woman's lover is described as grotesquely ugly, here attributed to his scabrous body:

"A very poor man who was covered with scabs was liked by nobody because of his disease. But the morning star, a woman who lives in the sky and who uses two mortars to pound algarroba, felt sorry for him, descended to the earth, and carried him to the sky."[182]

According to the Toba variant, Star Woman led her scabrous husband to a garden whereupon she transformed him "into a handsome young man."[183] The magical transformation of the wretched mortal at the hands of Venus offers a striking thematic parallel to Phaon's dramatic metamorphosis at the hands of

---

[176] *Ibid.*, p. 257.
[177] *Ibid.*, p. 261.
[178] J. Wilbert & K. Simoneau, *Folk Literature of the Ge Indians* (Los Angeles, 1978), p. 195.
[179] *Ibid.*
[180] J. Wilbert & K. Simoneau, *Folk Literature of the Toba Indians, Vol. 1* (Los Angeles, 1982), p. 55.
[181] *Ibid.*, p. 181.
[182] *Ibid.*, pp. 61-62.
[183] *Ibid.*, p. 56. See also A. Métraux, *op. cit.*, p. 44 where it is reported: "Upon nearing the garden, she transformed her husband into a handsome young man."

Aphrodite. Indeed, the fact that the planet Venus, as Star Woman, is credited with beautifying her scabrous paramour constitutes compelling circumstantial evidence that Aphrodite personifies Venus in her interactions with Phaon. This in itself is an important finding, one with profound and far-reaching implications for the history of Greek myth and religion.

That said, there is every reason to believe that the Star Woman myth has a great deal more to tell us. A number of parallels can be drawn between this myth and the Sumerian traditions describing the marriage of Inanna and Dumuzi. Thus it is that, in the sacred marriage rite, a prominent episode finds the mortal king achieving divinity—"stardom" in a quite literal sense—upon marrying the planet Venus. This is evidenced by the fact that early kings who performed the rite had the determinative denoting divinity appended to their names and received divine honors after their death.[184] The "deification" of the king, in our view, has its mythological prototype in Dumuzi's catasterization or apotheosis upon marrying Venus (here it will be remembered that, in BM 88348, the mortal hero is installed as a star alongside Inanna/Venus).[185]

A catasterization of the mortal hero is also attested in several versions of the Star Woman myth. According to one narrative, the mortal paramour is suddenly taken up to heaven to live alongside Venus: Hence it is reported of the (im)mortal hero that "now he is a star beside her in the heavens."[186]

The Sumerian text BM 96739, quoted earlier, likewise sheds additional light on the Star Woman myth. Recall here that the mortal hero (=the Sumerian king) is imbued with the luminous splendor or "glory" of the planet-goddess:

"Oh Inanna, a husband worthy of your splendor has been granted to you…You, oh mistress, you have handed over to him your power as is due to a king, and Ama-ušumgal-anna causes a radiant brilliance to burst out for you."[187]

It is our opinion that this mysterious episode provides the historical prototype and logical rationale for understanding the archaic myth of Star Woman. As Inanna/Venus was believed to confer power or "splendor" on Dumuzi so, too, does Star Woman/Venus confer youth and beauty on her wretched paramour.

---

[184] E. van Buren, "The Sacred Marriage in Early Times in Mesopotamia, Part II," *op. cit.*, p. 52. See also the discussion in H. Frankfort, *Kingship and the Gods* (Chicago, 1948), pp. 296-297.

[185] Pirjo Lapinkivi, *op. cit.*, p. 27 writes: "The ascension of Dumuzi to heaven and his being stationed there as a star can also be understood as a consequence of the union." See also W. von Soden, *The Ancient Orient* (Grand Rapids, 1994), p. 68.

[186] C. Nimuendajú, "Šerente Tales," *Journal of American Folklore* 57 (1944), p. 184.

[187] Quoted from F. Bruschweiler, *Inanna. La déesse triomphante et vaincue dans la cosmologie sumérienne* (Leuven, 1988), p. 110.

In each case a sexual union with a Star Woman has a dramatically transforming effect on the mortal hero, and in each case it is the Star's luminous efflux that transfigures him. Thus, Dumuzi is empowered and shines brilliantly as a result of his marriage with Inanna/Venus. In the Star Woman myth, similarly, the previously ugly mortal is transformed into a beautiful youth and "shines" with luminous splendor as a result of sexual union with Venus. A narrative from South America, cited earlier, captures the essence of this extraordinary effusion of luminous splendor characterizing "union" with Star Woman:

"When she lay down with him there was a light emanating from her, illuminating everything...The light, or maybe her beauty, was transmitted to her parents-in-law and to the other people who were there every night ...By now the other people and some girls were very envious of the young man's family because they looked so fair and beautiful. The man used to be dark and ugly, but when he slept with Iozly every night he grew better and better looking until he was fair and handsome, with smooth, fair hair."[188]

To summarize our findings in this chapter: However this imagery is to be understood from an astronomical standpoint, it is evident that the key to a proper interpretation is to understand Dumuzi in his original context—i.e., as a planetary body set alongside Venus. Inanna's empowerment of Dumuzi/Mars is best understood as having reference to a spectacular conjunction between two planets, one in which Inanna/Venus was seen to envelop or infuse Dumuzi/Mars with her luminous splendor.

---

[188] J. Wilbert & K. Simoneau, *Folk Literature of the Chamacoco Indians* (Los Angeles, 1987), pp. 85-86.

6. Aphrodite and Adonis

"Bethlehem which is now ours, the most august place in the universe… was shaded by the sacred wood of Tammuz, that is, Adonis. And in the grotto where the newborn Christ once cried, there were tears for the lover of Venus."[189]

The testimony of St. Jerome, taken together with that of other ancient writers, attests to Adonis's former exalted status. What, then, do we know about Aphrodite's celebrated paramour?

According to Panyassis (early 5th century BCE), the newborn Adonis was so beautiful that Aphrodite jealously hid him away in a coffin. After handing him over to Persephone for safekeeping, Aphrodite was subsequently heartbroken upon learning that the goddess of the underworld refused to give him up. Here is the account as preserved by Apollodorus:

"Struck by his beauty, Aphrodite, in secret from the gods, hid him in a chest while he was still a little child, and entrusted him to Persephone. But when Persephone caught sight of him, she refused to give him back. The matter was submitted to the judgment of Zeus; and dividing the year into three parts, he decreed that Adonis should spend a third of the year by himself, a third with Persephone, and the remaining third with Aphrodite…Later, however, while he was hunting, Adonis was wounded by a boar and died."[190]

---

[189] St. Jerome, Letters 58, 3 as translated by R. Turcan, *The Cults of the Roman Empire* (Oxford, 1996), p. 148.
[190] Apollodorus, *The Library* III:14:4 as translated in R. Hand ed., *The Library of Greek*

Although there are conflicting reports about the precise manner of the god's death, it is agreed that he died young and under tragic circumstances. According to one version of the myth, Aphrodite is said to have leapt off the Leucadian rock out of grief for the beautiful youth.[191]

Aphrodite's passion for Adonis is attested as early as Sappho (ca. 600 BCE). In a fragment attributed to the famed poet of Lesbos, one finds an early reference to the ritual lamentations that distinguished the god's cult:

"He is dying, O Cytherean, the tender Adonis! What shall we do? Beat your breasts, young maidens, and tear your tunics!"[192]

Bion, writing in the mid-second century BCE, composed a lengthy poem recounting Adonis's tragic fate. In his account it is the goddess Aphrodite herself who bloodied her breasts while mourning the youth's death:

"But Aphrodite, having let down her hair, rushes through the woods mourning, unbraided, unsandalled; and the thorns cut her as she goes and pluck sacred blood. Shrilly wailing, through long winding dells she wanders, crying out the Assyrian cry, calling her consort and boy. Around her floated her dark robe at her navel; her chest was made scarlet by her hands; the breasts below, snowy before, grew crimson for Adonis."[193]

The wailing rites alluded to by Sappho and Bion betray the telltale influence of Dumuzi's cult, whose proverbial lamentations are first attested in Mari during the Old Babylonian period (ca. 1800-1600 BCE) but undoubtedly go back much further still, likely to the dawn of civilization itself. Lamentations for the god are most familiar from the testimony of Ezekiel, who wrote as follows of the abominations then prevailing at Jerusalem: "Then he brought me to the door of the gate of the Lord's house which was toward the north; and, behold, there sat women weeping for Tammuz."[194] Such rites proved very difficult to extinguish and were still being performed by the Sabean women of Harran as late as the tenth century CE.[195]

The Adonis-myth formed the subject of several Greek rituals during the fifth and fourth centuries BCE.[196] Women were the primary participants in

---

Mythology (Oxford, 1997), pp. 131-132.

[191] See the discussion in L. Farnell, *The Cults of the Greek States, Vol. II* (New Rochelle, 1977), p. 650.

[192] Fragment 152 as translated in R. Turcan, *op. cit.*, p. 144.

[193] *Epitaph on Adonis*, lines 19-27 as quoted in J. Reed, *Bion of Smyrna* (Cambridge, 1997), pp. 123-125.

[194] *Ezekiel* 8:14.

[195] B. Alster, "Tammuz," in K. van der Toorn et al eds., *Dictionary of Deities and Demons in the Bible* (Leiden, 1995), col. 1569.

[196] S. Ribichini, "Adonis," in K. van der Toorn et al eds., *Dictionary of Deities and Demons*

the rites in question, known as Adonia, which were typically celebrated on rooftops, thereby emulating the Oriental custom. Interestingly, ladders formed a conspicuous element in the god's cult:

"According to textual evidence, Adonis rites were performed on the roofs of houses. The iconography contains some striking scenes in which ladders are outstanding features."[197]

In Athens, the Adonia featured the parading forth of the god's body and its burial, followed by a period of general licentiousness marked by drinking and dancing.[198] Further details, unfortunately, are lacking with regards to the precise order and content of the Attic rites. Walter Burkert emphasized the link with the ancient Near East in his summary of Adonis's cult in ancient Greece:

"There remain enough lacunae and uncertainties in our knowledge. Still we can feel confident as to the general outline: the yearly festival of weeping for Tammuz spread from Mesopotamia to Syria to Palestine, and thence, with the name 'Adonis,' to Greece. At Jerusalem, as still in fifth-century Athens, this is not an established state festival, but an unofficial ceremony spontaneously performed by women, and viewed with suspicion by the dominant male."[199]

Outside of Greece proper, there are indications that the cult of Adonis was once widely disseminated throughout the Mediterranean region. In Rome, as in Athens, Adonia were celebrated. Numerous Roman murals, according to Robert Turcan, show "Adonis being carried away by Venus."[200] The love of Adonis and Aphrodite was also a familiar subject on Etruscan mirrors from the fourth century BCE.

Adonis was especially popular at Byblos, a Phoenician stronghold of great antiquity.[201] Indeed, there is much reason to believe that Adonis was Astarte's youthful consort at Byblos. An eyewitness to the rites practiced there—Lucian (2nd Century CE)—reported that the god experienced a resurrection:

---

in the Bible (Leiden, 1995), cols. 12-17.

[197] T. Mettinger, *The Riddle of Resurrection* (Stockholm, 2001), p. 127. On the iconography of the ladder in the Adonia, see R. Rosenzweig, *Worshipping Aphrodite* (Ann Arbor, 2004), pp. 63-68.

[198] F. Graf, "Aphrodite," in K. van der Toorn et al eds., *Dictionary of Deities and Demons in the Bible* (Leiden, 1995), col. 120.

[199] W. Burkert, *Structure and History in Greek Mythology and Ritual* (Berkeley, 1979), p. 107.

[200] R. Turcan, *The Cults of the Roman Empire* (Oxford, 1996), p. 145.

[201] Vase fragments from the Fifth Dynasty reign of Unas have been found at Byblos. See W. Stevenson Smith, "The Old Kingdom in Egypt...," in I. Edwards et al eds., *The Cambridge Ancient History* Vol. 1:2 (Cambridge, 1971), p. 189.

> *"As a memorial of his suffering each year they beat their breasts, mourn and celebrate the rites. Throughout the land they perform solemn lamentations. When they cease their breast-beating and weeping, they first sacrifice to Adonis as if to a dead person, but then, on the next day, they proclaim that he lives and send him into the air."*[202]

Jerome, writing two centuries later than Lucian, provides complementary testimony in favor of a rite of resurrection involving Adonis. Witness his commentary on Ezekiel:

> *"What we have rendered as Adonis, the Hebrew and Syrian languages denote as Tammuz. According to a pagan tale, Venus's lover, a very beautiful youth, is killed...After this, he is said to have risen...There is an annual celebration of his feast, in which women bewail him as dead, and then he is praised in song when he returns to life...The same pagans interpret, in a subtle manner, the poets' narratives of a similar kind, narratives about shameful things: they understand the sequence of wailing and joy as referring to the death and resurrection of Adonis. They take his death to be shown by the seeds that die in the earth, and his resurrection by the crops in which the dead seeds are born."*[203]

Cyril of Alexandria, writing in the 5th century of the current era, commented on the Adonis rites then being celebrated in his native city. Cyril's disdain for the Greek practice is everywhere apparent:

> *"They pretended to unite in weeping and lamentations with Aphrodite when she was mourning Adonis's death. Then, when she reappeared from the Netherworld and announced that she had found the one she had been looking for, [they pretended] to unite in rejoicing and jumping [for joy]. And even today this comedy is still being performed in the temples of Alexandria."*[204]

In addition to the ritual lamentations, there is evidence that the Adonia featured a *hieros gamos* between the youthful hero and Aphrodite/Astarte. Such was the case in the rites practiced in Ptolemaic Alexandria, according to Theocritus (4th/5th century CE). The Alexandrian rites have been summarized by Sergio Ribichini as follows:

> *"The first day the participants celebrated the union between the two lovers, represented in the course of a banquet under a kiosk of dill stems and surrounded by fruits, delightful gardens, pots of perfumes and a big variety of cakes. On the second day the epithalamium gave*

---

[202] *De Dea Syria* 6.
[203] *Explanations in Ezekiel* III, 8, 14 as quoted from T. Mettinger, *op. cit.*, p. 130.
[204] *Isaiam* 18:1-2 as quoted in T. Mettinger, *op. cit.*, p. 123.

way to a lament as the worshippers gathered for a funeral procession to carry the image of Adonis to the seashore."[205]

The sacred marriage between Aphrodite and Adonis can't help but recall the *hieros gamos* associated with the Sumerian Inanna and Dumuzi. Inasmuch as the Sumerian rite had its origin and *raison d'être* in ancient conceptions associated with the planet Venus, one is naturally led to suspect that similar conceptions informed the aboriginal cult of Aphrodite and Adonis.

That there was a celestial dimension to the Adonis myth is also evidenced by the fact that his rites were typically celebrated on rooftops. It was on rooftops, after all, that astronomical observations and offerings were commonly made throughout the ancient Near East.[206] Jeremiah's testimony is especially instructive in this regard:

"And the house of Jerusalem, and the houses of kings of Judah, shall be defiled as the place of Tophet, because of all the houses upon whose roofs they have burned incense unto all the host of heaven, and have poured out drink offerings unto other gods."[207]

The Judaic rites are reminiscent of practices described in the sacred marriage hymn of Iddin-Dagan from more than a thousand years earlier. There one reads that incense was offered to the planet Venus on the rooftops:

"Everybody hastens to holy Inana. For my lady in the midst of heaven the best of everything is prepared (?). In the pure places of the plain, at its good places, on the roofs, on the rooftops, the rooftops of dwellings (?), in the sanctuaries (?) of mankind, incense offerings like a forest of aromatic cedars are transmitted to her."[208]

The Adonis rites are significant not only for the details they provide with respect to the celestial context of the sacred marriage rite but for the light they shed on dying gods in general and the myth of Dumuzi in particular. How, then, are we to understand the god behind these curious rites? As Ribichini points out with reference to the cult at Byblos, the name Adonis is most likely an epithet of a great god:

"He must indeed have been a god of high rank. It is probable that the cult of Adonis in Byblos continued the worship of a Phoenician 'Baal', conceived as a dying and rising god. This god was not merely a spring deity or vegetation spirit, as Frazer believed, but an important city god comparable to Melqart in Tyre and Eshmun in Sidon."[209]

---

[205] S. Ribichini, *op. cit.*, col. 13.
[206] M. Weinfeld, "The Worship of Molech and of the Queen of Heaven and its Background," *Ugarit-Forschungen* 4 (1972), pp. 151-154. See also *Zephaniah* 1:5.
[207] *Jeremiah* 19:13.
[208] Lines 142ff. in J. Black et al, *op. cit.*, p. 266.
[209] S. Ribichini, *op. cit.*, col. 14.

Melqart, in fact, was addressed as *Adon*, "my lord," the epithet from which derives the name of Adonis.[210] At Tyre, Melqart was the beloved consort of Astarte/Venus, thereby occupying a position similar to that of Adonis at Byblos.[211] Early on identified with Nergal, Melqart is best understood as a personification of the planet Mars,[212] and therefore it is interesting to find that Tyre was renowned for its worship of the red planet.[213] One Arabic author, writing in the first decades of the 14th century CE, offered direct testimony on the matter: "Among [the temples that were found] in the city of [Tyre], near the waterside, a temple of Mars."[214]

The author in question—Al-Dimashqi—goes on to state that Tammuz himself was to be identified with the planet Mars: "The Sabaeans contended that [Jerusalem] had been built before Solomon, peace be on him, and that the city had a temple of Mars where an idol called Tammuz was found."[215] How or from what sources Al-Dimashqi derived this information is not clear. That said, the fact that the Greek astronomer Ptolemy identified Adonis with the red planet offers additional support for Al-Dimashqi's claim. Thus, in a discussion of the inhabitants of Syria in his *Tetrabiblos*, he wrote as follows:

"Those who live in these countries generally worship Venus as the mother of the gods, calling her by various local names, and Mars as Adonis, to whom again they give other names, and they celebrate in their honour certain mysteries accompanied by lamentations."[216]

Ptolemy's primary works stem from the first half of the second century CE. The fact that he wrote from Alexandria, a hotbed of Adonis-worship, suggests that Greece's greatest astronomer was uniquely qualified to comment on the possible astronomical aspects of the Adonis-myth.

---

[210] KAI 47.
[211] T. Mettinger, *The Riddle of Resurrection* (Stockholm, 2001), p. 126.
[212] E. Cochrane, *Martian Metamorphoses* (Ames, 1997), pp. 42-48. See also S. Dalley, *Myths from Mesopotamia* (Oxford, 1989), p. 164: "The name of Melqart, chief god of Tyre, is a Phoenician translation of the Sumerian name Nergal, and they are thus very closely assimilated."
[213] H. Seyrig, "Antiquités Syriennes," *Syria* 64 (1944-45), p. 70, cites Saleh ibn Yahya for the worship of Mars in Tyre.
[214] Quoted from J. Hjärpe, *Analyse Critique des Traditions Arabes sur les Sabéens Harraniens* (Uppsala, 1972), p. 75. Note: I am indebted to Rens van der Sluijs for this reference and the translation thereof.
[215] *Ibid.*
[216] II, 3, 68. F. Robbins, translator, *Ptolemy: Tetrabiblos* (Cambridge, 1998), pp. 146-149.

7. Aphrodite: Star of Lamentation

> *"Children, Kypris is not Kypris alone, but she is called by many names. She is Hades, she is immortal life, she is raving madness, she is unmixed desire, she is lamentation; in her is all activity, all tranquility, all that leads to violence. For she sinks into the vitals of all that have life; who is not greedy for that goddess?"*[217]

In addition to her myriad other functions, Aphrodite offers an archetypal example of the mourning goddess. This idea is most obvious in the traditions surrounding Adonis, a god whose rituals featured ceremonial wailing and the singing of dirges to the accompaniment of shrill pipe music.[218] In the account of Bion of Smyrna, a poet of the late second century B.C.E., Aphrodite herself is said to have unbound her hair and embarked upon a period of wandering in the wake of Adonis's death:

> *"But Aphrodite, having let down her hair, rushes through the woods mourning, unbraided, unsandalled; and the thorns cut her as she goes and pluck sacred blood. Shrilly wailing, through long winding dells she wanders, crying out the Assyrian cry, calling her consort and boy. Around her floated her dark robe at her navel; her chest was made scarlet by her hands; the breasts below, snowy before, grew crimson for Adonis."*[219]

---

[217] From a fragment attributed to Sophocles, as quoted in Stephanie Budin, *The Origin of Aphrodite* (Bethesda, 2003), p. 18.

[218] L. Farnell, *op. cit.*, p. 637, observes: "We meet also with ceremonies of mourning and sadness in the worship of Leucothea at Thebes, and perhaps in Crete, as we find them elsewhere in the worship of Aphrodite."

[219] *Epitaph on Adonis*, lines 19-27 as quoted in J. Reed, *Bion of Smyrna* (Cambridge, 1997), pp. 123-125.

That Aphrodite's lamentations have some reference to the planet Venus receives a measure of support from Babylon, where Ishtar/Venus was known as the "star of lamentation."[220] This is indeed a puzzling epithet: What possible relation could there be between a distant planet and ancient mourning rites?

A survey of ancient goddesses will show that many were represented as great mourners. Inanna's lamentations in the wake of Dumuzi's death are representative in this regard. No ordinary lamentations, Inanna's are said to have shaken the very foundations of heaven:

"She of lament, she of lament, struck up a lament. The hierodule, she of lament, she of lament struck up a lament. The hierodule of heaven, Inanna, the devastatrix of the mountain, the lady of Hursagkalama, she who causes the heavens to rumble, the lady of the Eturkalama, she who shakes the earth...she of lament, she of lament (struck up a lament)."[221]

The destruction wrought by the raging mourning goddess was never wholly forgotten and figures prominently in the legendary accounts surrounding great kings. In a Sumerian hymn dating to the third millennium, the death of Ur-Namma (ca. 2100 BCE) is met with bitter wailing on the part of Inanna. In that context the following passage occurs: "Then Inana, the fierce storm...made the heavens tremble, made the earth shake."[222]

Anat's lamentations on behalf of Baal recall those of Inanna/Venus and were much celebrated in Canaanite tradition.[223] Witness the following passage:

"Then Anat went to and fro and scoured every mountain to the heart of the earth...She came upon Baal, fallen to earth. She covered her loins with sackcloth;...she scraped (her) skin with a stone...She gashed her cheeks and chin."[224]

A similar pattern is evident in Egyptian myth. A hymn to Osiris from the middle of the second millennium BCE offers a general summary of this archaic mythological theme:

"Isis, the powerful, the protector of her brother [i.e., Osiris], who searched for him without wearying, who traversed this land in grief and did not rest until she had found him, ...who sent up a cry, the mourning-woman of her brother."[225]

---

[220] F. Stephens, "Prayer of Lamentation to Ishtar," in J. Pritchard ed., *Ancient Near Eastern Texts* (Princeton, 1969), p. 384.

[221] M. Cohen, *Sumerian Hymnology* (Cincinnati, 1981), p. 148.

[222] Lines 204-205 from "The death of Ur-Namma (Ur-Namma A)," ETCSL. See also the discussion in E. Flückiger-Hawker, *Urnamma of Ur in Sumerian Literary Tradition* (Göttingen, 1999), pp. 86-87.

[223] N. Walls, *The Goddess Anat in Ugaritic Cult* (Atlanta, 1992), p. 67.

[224] *Ibid.*, pp. 68-69.

[225] J. Assmann, *The Search for God in Ancient Egypt* (Ithaca, 2001), pp. 145-147.

Analogous traditions surround the goddess Freyja, long acknowledged to be a Norse counterpart to the Latin goddess Venus. Thus, Snorri wrote that "Freyja cried (tears of) gold for Óðr."[226] As Robert Briffault recognized many years ago, Freyja's lamentations conform to a universal pattern:

"Freija was expressly a wanderer. Like Isis in search of Osiris, like Io and innumerable other goddesses, she wanders disconsolate in search of Odhr, or Odin."[227]

The same idea is attested in the New World. Thus, the Aztec goddess Itzpapalotl is said to have "wandered off—combing her hair, painting her face, and lamenting the loss of Arrow Fish."[228]

It is the Phrygian Cybele, perhaps, who offers the most interesting example of the lamenting goddess. According to Diodorus, the goddess wandered the world with disheveled hair while mourning the death of Attis.[229] Significantly, Cybele was identified with Aphrodite.[230]

There is good reason to believe that Diodorus's euhemerizing account preserves information of profound significance, as the mourning goddess's penchant for wandering around with wildly flowing hair forms a recurring theme in ancient myth. The Greek Electra, for example, is said to have loosed her hair and streamed across heaven as a comet while lamenting the destruction of Troy. Hyginus recounts Electra's flight as follows:

"But after the conquest of Troy and the annihilation of its descendants, ... overwhelmed by pain she separated from her sisters and settled in the circle named arctic, and over long periods she would be seen lamenting, her hair streaming. That brought her the name of comet."[231]

The lamenting Aphrodite, wandering about with streaming hair, naturally recalls Hyginus's description of the comet Electra. Is it possible that the lamenting planet-goddess presented a comet-like form? Certainly it is understandable that a planet displaying "disheveled hair" might be likened to a comet.

There is a wealth of circumstantial evidence, in fact, that the ancient traditions of lamenting goddesses have reference to a comet-like apparition. In the Sumerian hymn "Dumuzi's Dream," for example, Dumuzi's sister

---

[226] B. M. Näsström, "Freyja—A Goddess with Many Names," in S. Billington & M. Green eds., *The Concept of the Goddess* (London, 1996), p. 68.
[227] R. Briffault, *The Mothers, Vol. 3* (New York, 1927), p. 66.
[228] B. Brundage, *The Fifth Sun: Aztec Gods, Aztec World* (Austin, 1983), p. 171.
[229] *Library* 3:57-59.
[230] L. Farnell, *op. cit.*, pp. 633, 641.
[231] *De Astronomia*, as translated by Milad Doueihi in C. Sagan & A. Druyan, *Comet* (New York, 1985), p. 18.

Geshtinanna announces while lamenting his death that "my hair will whirl in heaven for you."[232] That the goddess's lamentations had reference to something actually seen in the sky is supported by a subsequent passage in the same hymn:

"Gestinanna cried toward heaven, cried toward earth. (Her) cries covered the horizon completely like a cloth and were spread out like linen."[233]

Ancient chroniclers also preserved memory of the planet-goddess's cometary form. Thus, various early Christian writers described a Phoenician ritual at Aphaca (Syria) in which the goddess Astarte was represented as a falling star.[234] Michael Astour summarized the ritual in question as follows: "It was believed that once a year the goddess descended into the pool as a fiery falling star, or that on solemn feast days, when people assembled in the shrine, a fire-globe was lit in the vicinity of the temple and probably rolled into the pool."[235]

Astarte herself, like the Greek Aphrodite, was explicitly identified with the planet Venus. It is no surprise, then, to find that the ancients recognized a fundamental affinity between the two goddesses. Witness the following statement of Philo: "The Phoenicians say that Astarte is Aphrodite."[236]

Aphrodite herself was evidently represented as a falling star at Paphos on Cyprus, being depicted by a large dark rock. Modern scholars have been inclined to identify the stone as meteoritic in nature:

"A large cone-shaped black stone was worshipped in the ancient temple of Aphrodite at Paphos, Cyprus. This temple, and the stone it contained, was celebrated on many Roman coins."[237]

As the "star of lamentation" was conceptualized as female in form so, too, did women feature prominently in ancient mourning rites.[238] Robert Briffault offered the following observation on this peculiarity of ancient ritual: "Those

---

[232] B. Alster, *Dumuzi's Dream* (Copenhagen, 1972), p. 61.
[233] *Ibid.*, p. 81.
[234] See here the discussion in J. Frazer, *Adonis, Attis, Osiris* (New York, 1961), p. 259.
[235] M. Astour, *Hellenosemitica* (Leiden, 1967), pp. 115-116, citing Sozomenos, *Historia Ecclesiastica* II:5; Zosimos, *Histories* I:58.
[236] Fragment 2, D32. See here H. Attridge & R. Oden, "Philo of Byblos: The Phoenician History," in *The Catholic Biblical Quarterly* 9 (1981), p. 55.
[237] M. D'Orazio, "Meteorite records in the ancient Greek and Latin Literature," in L. Piccardi & W. Masse, *Myth and Geology* (London, 2007), p. 220. See also V. Kenaan, "Aphrodite: The Goddess of Appearances," in A. Smith & S. Pickup eds., *Brill's Companion to Aphrodite* (Leiden, 2010), p. 29.
[238] M. Alexiou, *The Ritual Lament in Greek Tradition* (Cambridge, 1974), p. 212 writes: "The prominence of women in funeral lamentation is attested from earliest times in archaeology, epigraphy and literature."

rites and 'lamentations' are throughout the primitive society performed by women."[239] And as the "star of lamentation" typically displayed disheveled hair when mourning the death of her beloved consort so, too, are mourning rites around the globe often performed by women whose hair is purposefully loosened in order to appear disheveled. Arab mourners, for example, were described as follows by one scholar: "Then our women bewail (the dead) with voices, hoarse with weeping…with disheveled hair."[240] In the *Mahabharata*, mourning is signified by women wearing their hair loose.[241] Ancient Egyptian monuments likewise show women mourners with disheveled hair (see figure 1).[242] Given these widespread customs and the general belief that disheveled hair was a sign of mourning, it is doubtless no accident that various words for "mourning" in the Egyptian hieroglyphic language have the hair-sign as a determinative— ⌐ .[243]

Figure 1

Yet the question arises: If the Queen of Heaven is to be identified with the planet Venus, as the evidence demands, why would she be described with comet-like attributes? The answer to this question stands to revolutionize our understanding of the solar system's recent history.

---

[239] R. Briffault, *The Mothers* (New York, 1963), p. 173.
[240] A. Wensinck, *Some Semitic Rites of Mourning and Religion: Studies on Their Origin and Mutual Relation* (Amsterdam, 1917), p. 50.
[241] 2.71.18-20.
[242] In a death scene from a tomb at Saqqara, for example. See the discussion in A. Burton, *Diodorus Siculus: Book One, A Commentary* (Leiden, 1972), pp. 211, 261.
[243] W. W. "Trauer," in *Lexikon der Ägyptologie, Vol. V* (Berlin, 1977), p. 744.

8. The Venus-Comet

"From the book of Marcus Varro, entitled Of the Race of the Roman People, I cite word for word the following instance: 'There occurred a remarkable celestial portent; for Castor records that, in the brilliant star Venus, called Vesperugo by Plautus, and the lovely Hesperus by Homer, there occurred so strange a prodigy, that it changed its colour, size, form, course, which never happened before nor since.'[244] *St. Augustine*

That the planet Venus once presented a "comet-like" appearance is surely among the most controversial claims advanced in *Worlds in Collision*, a book otherwise renowned for generating controversy. Yet as we have seen in previous chapters, there is a wealth of evidence which can be brought to bear in support of this proposition. In this chapter we will review the ancient terminology surrounding comets in order to determine whether or not a consistent link to Venus can be established.

A survey of ancient comet lore reveals a dozen or so recurring terms for these extraterrestrial agents, the most common of which are the following: (1) hair star; (2) tailed star; (3) bearded star; (4) torch star; (5) dragon star; and (6) smoking star. Now it is a remarkable fact, first documented by Velikovsky,[245] that the very same terminology was used to describe the planet Venus.

Hair-Star

Comets were known as "haired stars" throughout the ancient world, the word comet itself deriving from the Greek *kometes*, the "long-haired." The Tshi of Africa, for example, refer to a comet as a "hair star."[246] In the New

---
[244] *City of God* 21:8 as translated in M. Dods, *The Works of Aurelius Augustine, Vol. 2* (Edinburgh, 1881), p. 429.
[245] *Worlds in Collision* (New York, 1950), pp. 173-176.
[246] S. Lagercrantz, "Traditional Beliefs in Africa Concerning Meteors, Comets, and Shoot-

World as well native peoples likened a comet to a "star with hair," "hairy star," or "maned star," appellations which accord completely with the global language of the comet. Thus, in Yucatec Maya dictionaries comets are referred to as "maned" stars.[247]

That New World peoples employed virtually identical terminology to describe Venus has been documented in previous chapters. The Inca name for Venus was *chasca coyllur*, signifying the "star (*coyllur*) with tangled or disheveled hair."[248] The Cuicatec Indians from Central America still describe Venus as "the star like a hairy beast."[249]

In this connection, it is significant to note that the sacred iconography of various indigenous cultures represents Venus as a "hair-star." Thus, an Inca painting shows Chasca/Venus as an orb with "hair-like" filaments radiating out in all directions (see figure 1).[250] This conforms exactly with the Inca's description of Venus as the star with "shaggy" or "disheveled" hair.

Figure 1

Illustrations of Venus from Chile offer a similar form (see figure 2).[251] The Chilean image, in turn, bears comparison with illustrations of Venus on Siberian shaman drums (see figure 3).[252]

---

ing Stars," in *Festschrift für Ad. Jensen* (Munich, 1964), p. 322.
[247] W. Lamb, "Star Lore in the Yucatec Maya Dictionaries," in R. Williamson ed., *Archaeoastronomy in the Americas* (Los Altos, 1981), p. 237.
[248] W. Sullivan, *The Secret of the Incas* (New York, 1996), p. 87.
[249] E. Hunt, *The Transformation of the Hummingbird* (Ithaca, 1977), p. 141.
[250] Adapted from E. Krupp, "Phases of Venus," *Griffith Observer* 56:12 (1992), with permission of the author.
[251] Adapted from E. Krupp, "Phases of Venus," *Griffith Observer* 56:12 (1992), with permission of the author
[252] Adapted from E. Krupp, "Phases of Venus," *Griffith Observer* 56:12 (1992), with permis-

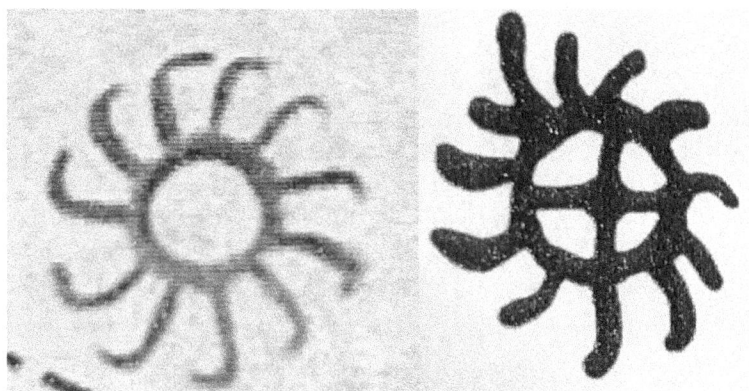

Figure 2   Figure 3

Similar ideas are attested in the Old World. The Latin scholar Varro, in a discussion of the planet Venus, noted that it was called *Iubar* because it was maned: "The morning-star is called *iubar*, because it has at the top a diffused light, just as a lion has on his head a *iuba* 'mane.'"[253] Yet the same Latin term was also used to describe a comet.

Varro's discussion is intriguing inasmuch as several other cultures compared the planet Venus to a lion. In Sumerian literature, as we documented in a previous chapter, the planet-goddess Inanna was described as the "Inana, great light, lioness of heaven."[254]

Lions are well-attested in the sacred iconography surrounding the Venus-goddesses. In the Babylonian cult of Ishtar, a popular motif finds lions being marked with a "hair-star" on their bodies, various authorities noting that the design "was a token of possession marking…animals [with it] as the property of Ishtar" (see figure 4).[255] The presence of a "hair-star" on the sacred animals of Ishtar/Venus seems especially fitting if that planet-goddess once presented a comet-like appearance.

---

sion of the author.
[253] *De lingua Latina* VII:76. See also J. Sammer, "An Ancient Name for Venus," *Kronos* VI:2 (1981), p. 61.
[254] Line 2 from "A hymn to Inana as Ninegala (Inana D)," *ETCSL*.
[255] E. van Buren, "An Additional Note on the Hair-Whirl," *Journal of Near Eastern Studies* 9 (1950), p. 55.

Figure 4

Most significant, perhaps, is the fact that Ishtar's star was occasionally depicted as "hairy" or "bushy," as on the cylinder seal depicted in figure 5.[256]

Figure 5

Tailed-Star

That comets were called "tailed" stars by numerous cultures is well-documented.[257] Such was the case amongst Germanic peoples, according to Grimm.[258] Comets were known as *fetia ave*, or "stars with tails," amongst the Polynesian Islanders.[259] The Tewa Indians of the Rio Grande called comets "tailed stars."[260] The Pomeroon Arawaks of South America describe a comet as a "star with tail."[261] Among the Maya, a comet was known as *uhe 4humil*, "star's tail."[262] The same idea is widespread throughout Africa.[263]

---

[256] Adapted from figure 317b in O. Keel & C. Uehlinger, *Gods, Goddesses, and Images of God in Ancient Israel* (Minneapolis, 1998), p. 324.

[257] B. Tedlock, "The Road of Light…" in A. Aveni ed., *The Sky and Mayan Literature* (Oxford, 1992), p. 28.

[258] J. Grimm, *Teutonic Mythology, Vol. 2* (Gloucester, 1976), p. 722.

[259] R. Williamson, *Religious and Cosmic Beliefs of Central Polynesia, Vol. 1* (Cambridge, 1933), p. 127.

[260] R. Williamson, *Living the Sky* (Norman, 1984), p. 189.

[261] W. Roth, "An Inquiry into the Animism and Folklore of the Guiana Indians," *Bureau of American Ethnology* 30 (1915), p. 259.

[262] D. Tedlock, *Popol Vuh* (New York, 1985), p. 348.

[263] S. Lagercrantz, "Traditional Beliefs in Africa Concerning Meteors, Comets, and Shoot-

Yet Venus was also described as a "tailed star." Recall the Yakut legend surrounding Venus, quoted earlier:

> "It [Solbon, the planet Venus] is said to be 'the daughter of the Devil and to have had a tail in the early days.' If it approaches the earth, it means destruction, storm, and frost, even in the summer; 'Saint Leontius, however, blessed her and thus her tail disappeared.'"[264]

Bearded-Star

The designation of comets as "bearded stars" goes back at least to Aristotle and continued well into modern times.[265] In light of this widespread terminology, it is intriguing to find that Assyro-Babylonian astronomical texts described the planet Venus as "bearded."[266] Such reports have generated a fair amount of commentary from scholars exploring these early texts. Morris Jastrow, citing a planetary omen from the time of Ashurbanapal (c. 668-626 BCE) referring to Venus's beard, offered the following explanation:

> "It is evident from this that the expression 'Venus has a beard' refers to some phenomenon connected with the appearance of the planet. In order, however, to remove all doubt as to the meaning of the phrase, the scribe has been careful enough to add an explanatory comment as follows: zikna zak-nu ziknu (or zakânu) na-ba-tu ba-'i-lat ni-bat, i.e., in the phrase zikna zak-nu, the term ziknu ('beard') means 'to shine', and the entire phrase therefore 'she shines strongly'."[267]

It is equally likely, however, that the gloss added by the ancient scribe represents simply his own best guess for what was, in reality, a very old idea attached to the planet-goddess, one no longer understood.[268] The specific context of the planet-goddess's beard in ancient myth and art suggests that this was, in fact, the case.

Vestiges of an anomalous "beard" are also present in the cults of other Venus-goddesses as well. The Cypriote Aphrodite was depicted as bearded,

---

ing Stars," in *Festschrift für Ad. Jensen* (Munich, 1964), p. 322.

[264] L. Mándoki, "Two Asiatic Sidereal Names," in V. Dioszegi ed., *Popular Beliefs and Folklore Traditions in Siberia* (Bloomington, 1968), p. 489.

[265] U. Dall'Olmo, "Latin Terminology Relating to Aurorae, Comets, Meteors, and Novae," *Journal for the History of Astronomy* 11 (1980), pp. 16-20.

[266] M. Jastrow, "The Bearded Venus," *Revue Archéologique* 17 (1911), pp. 271-298.

[267] *Ibid.*, p. 272.

[268] D. Brown, *Mesopotamian Planetary Astronomy-Astrology* (Groningen, 2000), p. 159 observes: "In certain cases the meanings of some technical terms in the protases [of astronomical omens] were lost, and in order to make sense of otherwise incomprehensible omens these were reinterpreted."

as was the Latin Venus.[269] Inasmuch as each of these goddesses represented the very ideal of beauty and femininity for their respective cultures, the presence of a beard is difficult to explain apart from their identification with the "female star."[270]

Torch-Star

Comets have been known as "torch stars" from time immemorial.[271] Yet as we have seen, it is common to find Venus-goddesses described as a "torch." Inanna/Venus, for example, was invoked as "the pure torch that flares in the sky, the heavenly light, shining bright like the day, the queen of heaven."[272] So, too, a leading epithet of Ishtar was "brilliant torch of heaven and earth."[273]

Scholars have typically sought to understand Inanna/Ishtar's "torch"-epithet by reference to Venus's brilliant appearance in the evening sky. Bob Forrest defended this position in his critique of Velikovsky:

"Velikovsky claims that Chaldean descriptions of the planet Venus are not consistent with the Venus we see today—for example, 'bright torch of heaven' and 'diamond that shines like the sun'...But when Venus is at its brightest in the early evening sky it is a beautiful sight, and it requires very little imagination to see it as a 'bright torch of heaven'."[274]

Although perfectly reasonable as a first attempt to account for the imagery in question, Forrest's interpretation fails to convince because it ignores the destructive behavior associated with the heavenly "torch" (Sumerian *izi-gar*). Indeed, it is the catastrophic context of Inanna's incendiary rampaging which rules out the conventional explanation. Witness the following passages in which the planet-goddess is likened to fire (Sumerian *izi*) and said to rain fire from heaven:

"Celestial luminary, you're like the fire! Verily you [shake?] the earth. Hierodule Inanna, celestial luminary, you are like the fire!"[275]

---

[269] Macrobius, *Saturnalia* 3:8.2. See also L. Farnell, *The Cults of the Greek States, Vol. 2* (New Rochelle, 1977), p. 628.

[270] This remains true even if one chooses to explain the goddesses' beards by diffusion from the cult of Ishtar—a distinct possibility—as such borrowing would be most unlikely to occur in the absence of a clear recognition of their planetary nature.

[271] U. Dall'Olmo, *op. cit.*, pp. 16-20.

[272] T. Jacobsen, *The Treasures of Darkness* (New Haven, 1976), p. 139.

[273] A. Sjöberg, "in-nin šà-gur$_4$-ra. A Hymn to the Goddess Inanna...," *Zeitschrift für Assyriologie* 65 (1976), p. 242.

[274] B. Forrest, *Velikovsky's Sources* (Santa Barbara, 1987), p. 23.

[275] M. Cohen, *Sumerian Hymnology* (Cincinnati, 1981), p. 130.

"You are the celestial luminary blazing like fire upon the earth."[276]

The Akkadian Ishtar is described in analogous terminology. Especially significant is the fact that Ishtar's hierophany as "the torch of heaven" is indistinguishable from her warrior aspect:

"Planet for the warcry...Gushea [an epithet of Ishtar], whose mail is combat, clothed in chilling fear...At the thought of your name, heaven and the netherworld quake...Shining torch of heaven...Fiery glow that blazes against the enemy, who wreaks destruction on the fierce, Dancing One, Ishtar..."[277]

In hymn after hymn, Ishtar/Venus is described as a fire-spewing warrior whose terrifying rampages usher in widespread destruction and shake the very foundations of the world:

"I rain battle down like flames in the fighting, I make heaven and earth shake with my cries...I, Ishtar, am queen of heaven and earth. I am the queen...I constantly traverse heaven, then (?) I trample the earth, I destroy what remains of the inhabited world."[278]

Now I ask: Does any of this language make sense in terms of the current planet Venus? By what stretch of the imagination can Venus be said to rain fire or shake heaven and earth? Yet very similar descriptions of the warrior-goddess's terrifying onslaught can be found throughout the ancient world, as we have documented elsewhere.[279]

Although a detailed analysis of the language associated with Inanna/Ishtar as the "torch-star" would be impossible here, one last example will suffice to illustrate the remarkable coherence of the imagery involved, hitherto overlooked by conventional scholars. In a late Babylonian hymn, Inanna is described as follows: "May your torch, which spreads terror, flare up in the heart of heaven."[280] It will be noted that the "terror" associated with Venus—however we are to understand the Akkadian word *ša-lum-mat*—is said to flare up from the "heart of heaven." Yet as we have seen, such a position is impossible for Venus to assume given the current arrangement of the solar system. Most significant, however, is the fact that ancient scribes elsewhere compared the terrifying *ša-lum-mat* to the glow of a comet (*šallummu*):

---

[276] *Ibid.*, p. 134.
[277] B. Foster, *Before the Muses* (Bethesda, 1993), pp. 510-512.
[278] *Ibid.*, p. 77.
[279] E. Cochrane, "The Birth of Athena," *Aeon* 2:3 (1990), pp. 26-28.
[280] B. Hruška, "Das spätbabylonische Lehrgedicht 'Inanna's Erhöhung'," *Archiv Orientalni* 37 (1969), p. 492.

"If a UL (comet) that has a crest in front and a tail in back is seen and lights up the sky like a šallummu...A šallummu equals an awesome radiance [ša-lum-ma-tu], An awesome radiance (a comet) equals an awesome radiance [me-lam-mu]."[281]

As Chadwick pointed out in his commentary on the cuneiform text in question, the most likely reason for the comparison is that the terrifying sheen presented by comets paralleled the terrifying radiance elsewhere associated with Inanna's torch-like epiphany: "The glow of the *šallummu* is likened to the terrifying light that surrounded deities which were known as *šalummatu* and *melammu*."[282]

Inanna's fearsome splendor is a recurring point of emphasis in Sumerian literature.[283] Recall the passage quoted earlier: "Agitation, terror, fear, splendour, awe-inspiring sheen are yours, Inanna."[284] One can be certain that the author of this hymn was not describing the familiar Venus on a clear summer night. Rather, Enheduanna was describing a terrifying heaven-spanning form—the planet-goddess known as Inanna—one that had assumed a "fearsome splendor" akin to that otherwise associated with a great comet. In this passage, as throughout her corpus of hymns describing Inanna's destructive behavior, Enheduanna's testimony offers a perfect example of what Jacobsen called a confrontation with the numinous—"a *mysterium tremendum et fascinosum*, a confrontation with a 'Wholly Other' outside normal experience and indescribable in its terms: terrifying, ranging from sheer demonic dread through awe to sublime majesty; and fascinating, with irresistible attraction, demanding unconditional allegiance."[285]

Dragon-Star

Various cultures have viewed comets as dragons or serpents moving across the sky, spewing venomous fire.[286] The Aztecs, among others, referred to a

---

[281] R. Chadwick, "Identifying Comets and Meteors in Celestial Observation Literature," in H. Galter ed., *Die Rolle der Astronomie in den Kulteren Mesopotamiens* (Graz, 1993), pp. 173-174.

[282] *Ibid.*, p. 174.

[283] See the extensive and illuminating discussion in F. Bruschweiler, *op. cit.*, pp. 116-175.

[284] Å. Sjöberg, "in-nin šà-gur₄-ra. A Hymn to the Goddess Inanna...," *Zeitschrift für Assyriologie* 65 (1976), p. 195.

[285] T. Jacobsen, *The Treasures of Darkness* (New Haven, 1976), p. 3.

[286] In *Funk and Wagnall's Standard Dictionary of Folklore, Mythology, and Legend* (New York, 1949), p. 243 one reads: "Early conceptions of the comet saw in it a dragon or serpent."

comet as a "star serpent."[287] The astronomer Peter Brown observed that the aboriginal peoples of Mesoamerica represented comets "by the plumed serpent depicted in various forms."[288]

Similar ideas are attested in Old World cultures. Thus, an Anglo-Saxon Chronicle from 793 alludes to the disasters which followed in the wake of dragons appearing in the sky:

"Fierce, foreboding omens came over the land of Northumbria and wretchedly terrified the people. There were excessive whirlwinds, lightning storms and fiery dragons were seen flying in the sky. These signs were followed by great famine."[289]

If scholars have been hard-pressed to explain the Mesopotamian traditions of Venus as a warring "torch," they are at a complete loss for words when encountering descriptions of the planet as a fire-spewing dragon. In a remarkable passage from *The Exaltation of Inanna*, the planet-goddess is described as a dragon raining fire from the heavens:

"Like a dragon you have deposited venom on the land, When you roar at the earth like Thunder, no vegetation can stand up to you. A flood descending from its mountain, Oh foremost one, you are the Inanna of heaven and earth! Raining the fanned fire down upon the nation..."[290]

Since Venus does not currently present the appearance of a serpent-dragon raining fire, scholars have been at pains to divorce Inanna from the planet when attempting to understand such imagery. Thus, while admitting that a serpent-dragon would be a fine description of a brilliant comet, Bob Forrest opined that Inanna's appearance as a dragon had reference to the goddess as a personification of the earth![291] Such an interpretation is belied by the evidence, however, for the ancient texts leave no room for doubt that it is a celestial body (Venus) that is the subject of the ophidian imagery surrounding Inanna/Ishtar. Witness the following passage from a Sumerian temple-hymn:

"Inanna...the great dragon who speaks inimical words to the evil, ... Through her the firmament is made beautiful in the evening."[292]

The astronomers Clube and Napier, while conceding that cometary motifs inform the cult of Inanna, would follow Forrest in seeking to divorce the goddess from the planet otherwise associated with the Queen of Heaven.

---

[287] C. Burland, *The Aztecs* (London, 1980), p. 102.

[288] P. Brown, *Comets, Meteorites and Men* (New York, 1973), p. 18.

[289] Quoted from V. Clube & W. Napier, *The Cosmic Winter* (London, 1990), p. 19.

[290] W. Hallo & J. van Dijk, *The Exaltation of Inanna* (New Haven, 1968), p. 15.

[291] B. Forrest, *Velikovsky's Sources, Notes and Index Volume, Vol. 3* (1983), p. 227.

[292] Å. Sjöberg & E. Bergmann, *The Collection of the Sumerian Temple Hymns* (Locust Valley, 1969), p. 36.

Pointing to "new astronomical information," the authors offered the following reinterpretation:

> *"Inanna is usually interpreted by scholars as the planet Venus, and in the absence of new astronomical information this is no doubt the best that one could have done. However it seems that the settled, spectacular, celestial imagery of the goddess Inanna, a morning and evening object 'crowned with great horns' and associated with the omega symbolism of Hathor, is more compatible with the thesis that the goddess was a great comet in a short-period orbit."*[293]

But there is no evidence that Inanna was ever represented as a "great comet in a short-period orbit" during the 3000 years of Mesopotamian history. By resorting to such baseless speculations, Clube and Napier manage to avoid drawing the one conclusion that is most compatible with the full body of evidence: It was Venus itself which only recently presented a comet-like appearance.

"Smoking star"

The Chukchee of northeastern Siberia refer to comets as "smoking stars."[294] Among Polynesian Islanders, comets were known as *pusa-loa*, or "elongated smoke."[295] Various African peoples likewise compare the tail of a comet to "smoke." The Djaga, for example, view the tail as "smoke of the smallpox fire" and believe it portends a smallpox epidemic.[296]

Similar conceptions are attested in the New World. Among the various names for "comet" in Maya lore was *budz ek*, "smoke stars."[297] So, too, an Aztec name for comet was *citlalimpopoca*, or "smoking star."

In *Skywatchers of Ancient Mexico*, the astronomer Anthony Aveni presented a series of illustrations dating to the sixteenth century, some of which are accompanied by captions pertaining to comet lore. Several make reference to the impending death of a ruler or great chief and thus are of interest inasmuch as they conform with a universal belief surrounding comets. But it is another caption affixed to a comet-like object which is of utmost interest for, as Aveni reports, "the caption in 9h tells us that the star Venus is smoking."[298]

---

[293] *The Cosmic Winter* (London, 1990), p. 180.
[294] W. Bogoras, "The Folklore of Northeastern Asia, as compared with that of Northwestern America," *American Anthropologist* 4 (1902), p. 593.
[295] R. Williamson, *op. cit.*, p. 132.
[296] S. Lagercrantz, *op. cit.*, p. 322.
[297] W. Lamb, "Star Lore in the Yucatec Maya Dictionaries," in R. Williamson ed., *Archaeoastronomy in the Americas* (Austin, 1981), p. 237.
[298] A. Aveni, *Skywatchers of Ancient Mexico* (Austin, 1981), p. 27.

Puzzled by this report, Aveni goes on to speculate that "perhaps a cometary object appeared near the planet."[299] An alternative interpretation would take the report as it stands and consider the possibility that the Aztecs—rightly or wrongly—ascribed a comet-like nature to Venus.[300] Maya lore provides some support for this interpretation. Thus, a text in the so-called *Songs of Dzitbalche* describes Venus as a "smoking star."[301]

It is notable that analogous beliefs are attested in Europe around the time of the Renaissance, a period of intense interest in comets. Horatio Grassi, for example, a participant in a series of famous debates with Galileo inspired by the appearance of three comets in 1618, observed that until recently the ignorant mass "had considered Venus as a comet."[302]

By itself, of course, Grassi's report counts for little and might easily be dismissed as idle gossip or rank superstition. Yet when viewed in the light of the roughly contemporaneous testimony from the greatest astronomers of the New World, Grassi's statement takes on an added significance and might well be interpreted as a faint reminiscence of ancient traditions no longer understood.

Summary

A survey of the lexicons of various ancient cultures reveals that the terminology attached to comets finds a striking correspondence with that employed to describe the planet Venus. On this score the evidence from ancient language complements the testimony from ancient literature, which likewise describes the planet Venus with comet-like characteristics (disheveled hair, serpentine form, tailed body, bearded, agent of eclipses, etc).

In reflecting upon the terminology shared by Venus and comets, it is essential that one keep in mind the fundamental purpose of language. Language is a means of communicating information with regards to the nature and appearance of the world around us. In origin and function, individual nouns and adjectives serve to distinguish one natural object from another. Names and epithets applied to Venus, for example, ought to serve to distinguish that planet from other celestial bodies. Thus, Venus might be called the "white star" in order to distinguish it from Mars, the "red star." Communication would be hindered, and confusion would soon result, if the names and epithets of

---

[299] *Ibid.*
[300] This was the position defended by Immanuel Velikovsky in *Worlds in Collision* (New York, 1950), p. 173.
[301] M. Edmonson, "The Songs of Dzitbalche: A Literary Commentary," *Tlalocan* IX (1982), p. 183.
[302] H. Grassi, "An Astronomical Disputation on the Three Comets of the year 1618," in S. Drake & C. O'Malley eds., *The Controversy of the Comets of 1618* (Philadelphia, 1960), p. 7.

the respective planets were to be employed in a haphazard fashion, with little regard for observational reality. Hence the remarkable anomaly presented by the fact that Venus and comets share so much terminology in common.

It is true that there will always be room for some occasional overlapping of terminology, particularly when the respective celestial bodies share a fundamental attribute in common. Thus, it is easy to understand how Venus, like the Sun, might be compared to a brilliant gem gleaming in the sky. But such natural vicissitudes of language will never explain why Venus was described as a "tailed" or "dragon-like" star.

It is also to be expected that, with the continued evolution of a particular language, such factors as metaphor, mythical imagery, and sympathetic magic will come to play a certain role in the application of names and epithets to the various celestial bodies. Witness the modern names applied to Venus: Sister Planet, Cytherea, twin, etc. Yet such factors will never explain why Venus was described with comet-like terminology in the Old World as well as the New for the simple reason that it is exceedingly unlikely that the very same metaphor or "mythical" interpretation would occur to different cultures in the absence of an empirical basis in natural history. Thus we can reasonably conclude that the skywatchers of Mesopotamia, like the skywatchers of Mesoamerica, compared Venus to a serpent-dragon because that planet once presented the appearance of a fiery serpent spanning the skies.

9. Aphrodite: Warrior Goddess

"In order to understand Aphrodite, who is so close to Ares, we cannot ignore her links with virility, vital strength, violence, and the physical dimension."[303]

"As the lady, admired by the Land, the lone star, the Venus star, the lady elevated as high as the heaven, ascends above like a warrior, all the lands tremble before her."[304]

In ancient Greece Aphrodite was invoked as a warrior—hence the epithet *Areia*. As Fritz Graz has pointed out, this particular image of the goddess puzzled even the Greeks themselves: "The armed Aphrodite of Sparta challenged the wits of Hellenistic epigrammists and Roman students of rhetoric: for both, she was a puzzling paradox."[305] Yet the Spartan cult finds a precise parallel on the island of Cythera, where Aphrodite *Urania* was represented as armed. And the Cytherean cult, it will be remembered, was deemed to be especially archaic. Farnell's opinion on the matter seems perfectly justified: "We may believe that the cult of the armed Aphrodite belongs to the first period of her worship in Greece."[306]

How, then, are we to understand Aphrodite's role as a warrior? The Greek evidence is only of minimal help here, being relatively meager in nature, due in

---

[303] G. Pironti, "Rethinking Aphrodite as a Goddess at Work," in A. Smith & S. Pickup eds., *Brill's Companion to Aphrodite* (Leiden, 2010), p. 129.

[304] Lines 135-137 from "A *šir-namursaga* to Inana for Iddin-Dagan (Iddin-Dagan A)," *ETCSL*.

[305] F. Graz, "Women, War, and Warlike Divinities," in W. Eck et al eds., *Zeitschrift für Papyrologie und Epigraphik* (1984), p. 250.

[306] L. Farnell, *The Cults of the Greek States, Vol. II* (New Rochelle, 1977), p. 653.

no small part to the fact that by the time of our earliest testimony the goddess had become largely humanized and more than a little specialized. As Jane Harrison pointed out many years ago, there is a marked tendency in ancient Greece for originally multifaceted goddesses to become compartmentalized over the course of time. With respect to Aphrodite, Harrison offered the following observation:

> *"Another note of her late coming into Greece proper is that she is in Homer a departmental goddess, having for her sphere one human passion. The earlier forms of divinities are of larger import, they tend to be gods of all work. When the fusion of tribes and the influence of literature conjointly bring together a number of local divinities, perforce, if they are to hold together, they divide functions and attributes, i.e., become departmental."*[307]

On this aspect of Aphrodite's cult, as with so many others, comparative mythology is our surest guide. From ancient Sparta we turn to the testimony from ancient Mesopotamia.

A Sumerian kenning, attested already in the third millennium BCE, held that "battle is the dance of Inanna."[308] For a comparative mythologist kennings are a bit like fossilized teeth for a paleontologist: They offer invaluable clues with regards to prehistoric functional and structural relationships long after their original function had come to an end. In this case the kenning attests to Venus's archaic and inherent identification as a warrior-goddess, attested around the globe.

The first point to be made is that the warrior-goddess is intimately related to the lamenting goddess. If, in one verse, Inanna is described as a great warrior whose "raging" threatens to destroy heaven and earth, another verse describes her as a grieving mourner whose tortured wailings shake the foundations of the world:

> *"Devastatrix of the lands, you are lent wings by the storm...you fly about the nation. At the sound of you the lands bow down. Propelled on your own wings you peck away at the land. With a roaring storm you roar; with Thunder you continually thunder...To (the accompaniment of) the harp of sighs you give vent to a dirge."*[309]

Why exactly the planet-goddess Inanna would be represented as a raging warrior has received precious little attention from scholars. Jacobsen,

---

[307] J. Harrison, *Prologemena to the Study of Greek Religion* (New York, 1975), p. 308.

[308] Line 289 from "Enmerkar and the lord of Aratta," *ETCSL*. See also T. Jacobsen, *The Treasures of Darkness* (New Haven, 1976), p. 137.

[309] W. Hallo & J. van Dijk, *The Exaltation of Inanna* (New Haven, 1968), pp. 17-19.

in introducing the goddess's warrior aspect, remarks: "In the process of humanization, gods of rain and thunderstorms tended...to be envisaged as warriors riding their chariots into battle."[310] Why this should be the case is not explained and, in any case, is far from obvious.

Joan Westenholz sought to trace Inanna's warlike nature to her intimate association with the rites of kingship: "Her function as a bestower of kingship and protectress of the city-state of Uruk may have given rise to her warlike character, since kingship followed the fortune of arms."[311] Jimmy Roberts sought to explain Ishtar's warrior prowess as follows: "Since in early nomadic society the young women egged on the young warriors in battle with praise and taunts, she could also be seen as the personification of the rage of battle."[312]

Such hypotheses are distinguished by their superficial and ad hoc nature. In addition to ignoring the explicit catastrophic context of the planet-goddess's war-mongering, the remarkably rich and detailed imagery associated with her raging aspect is passed over in silence. The fact that Inanna is explicitly identified with the planet Venus *but also as a warrior* already at the dawn of history, while acknowledged, is treated as if it is irrelevant to the imagery in question.[313] Yet nothing could be further from the truth. Witness the opening lines of Iddin-Dagan's marriage hymn, wherein the planet-goddess is expressly described as a warrior who descends *from the sky*:

"I shall greet her who descends from above...I shall greet the great lady of heaven, Inana!...I shall greet the Mistress, the most awesome lady among the Anuna gods; the respected one who fills heaven and earth with her huge brilliance...Her descending is that of a warrior."[314]

That scholars have been inclined to divorce the goddess's warrior-aspect from the planet Venus is hardly surprising, for what could such imagery have to do with the planet known to modern astronomers? The familiar Venus, after all, typically displays a beautiful and benign appearance and never wars, rages, storms, laments, or otherwise displays a threatening persona. Indeed,

---

[310] T. Jacobsen, *The Treasures of Darkness* (New Haven, 1976), p. 137.

[311] J. Westenholz, "Goddesses of the Ancient Near East 3000-1000 BC," in L. Goodison & C. Morris eds., *Ancient Goddesses* (London, 1998), p. 73.

[312] J. Roberts, *The Earliest Semitic Pantheon* (Baltimore, 1972), p. 40.

[313] Thus, Edzard notes that the astral aspect of Inanna/Ishtar is frequently expressed together with the warlike aspect of the goddess. See D. O. Edzard, "Mesopotamien: Die Mythologie der Sumerer und Akkader," in *Wörterbuch der Mythologie*, ed. by H. Haussig (Stuttgart, 1962), p. 85. See also the discussion in H. Balz-Cochois, *Inanna* (Gütersloh, 1992), p. 46.

[314] Lines 1-18 as quoted from "A *šir-namursaga* to Inana for Iddin-Dagan," in J. Black et al, *The Literature of Ancient Sumer* (Oxford, 2004), p. 263. See also D. Reisman, "Iddin-Dagan's Sacred Marriage Hymn," *Journal of Cuneiform Studies* 25 (1973), pp. 186-191.

it is the striking incongruity between Inanna's dual personality as Venus *and* as a warrior which has led scholars to speak of a "coalescence" of originally disparate cults under the name of Inanna.[315]

Incongruous or not, great goddesses are represented as warriors around the globe. In addition to Inanna and Aphrodite, Anat,[316] Astarte,[317] al-'Uzza[318] Freya[319] and Hathor are all represented as fierce warriors. The traditions surrounding Ishtar are especially instructive here:

"I rain battle down like flames in the fighting, I make heaven and earth shake (?) with my cries, ...I, Ishtar, am queen of heaven and earth. I am the queen...I constantly traverse heaven, then (?) I trample the earth, I destroy what remains of the inhabited world."[320]

A complementary picture is offered by the following passage:

"O splendid lioness of the Igigi-gods, who renders furious gods submissive...great is your valor, O valiant Ishtar, Shining torch of heaven and earth, brilliance of all inhabited lands. Furious in irresistible onslaught, hero to the fight, Fiery glow that blazes against the enemy, who wreaks destruction on the fierce, Dancing one, Ishtar."[321]

The Akkadian word translated as "lioness" here is *Labbatu*. Interestingly enough, the very same epithet signifies a goddess of lamentation:

"A name of Ištar in god lists and an epithet in texts. The special reference of this name of the goddess is given as 'of lamentation' (ša lal-la-ra-te) in CT 24, 41, 83, but the basis of this interpretation is not clear."[322]

Why the concepts "lioness" and "of lamentation" should coalesce around the planet-goddess Ishtar has long baffled scholars. In this context it is intriguing to recall that the Latin scholar Marcus Varro noted of the planet Venus that it was called *Iubar* "because it is *iubata* 'maned'."[323]

---

[315] T. Jacobsen, *The Treasures of Darkness* (New Haven, 1976), p. 135.

[316] See the discussion in A. Eaton, *The Goddess Anat: The History of Her Cult, Her Mythology and Her Iconography* (New Haven, 1969), pp. 54-78.

[317] J. Westenholz, *op. cit.*, p. 79.

[318] H. Drijvers, "The Cult of Azizos and Monimos at Edessa," in C. J. Bleeker et al eds., *Ex Orbe Religionum* (Leiden, 1972), p. 364.

[319] E. O. Turville-Petre, *Myth and Religion of the North* (New York, 1964), p. 177.

[320] B. Foster, *Before the Muses: An Anthology of Akkadian Literature*, Vol. 1 (Bethesda, 1993), p. 74.

[321] *Ibid.*, 512.

[322] W. Lambert, "Labbatu" in E. Ebeling & B. Meissner, eds., *Reallexikon der Assyriologie*, Vol. 6 (Berlin, 1980-1983), p. 411.

[323] *De lingua latina* VI:6. See here the discussion in J. Sammer, "An Ancient Latin Name for Venus," *Kronos* 6:2 (Winter 1981), p. 61.

Although explicitly catastrophic in nature, scholars have virtually never considered the possibility that the vivid language describing the rampaging Inanna/Venus might have had reference to cataclysmic natural events.[324] Yet there is compelling evidence that the imagery in question was inspired by the planet Venus itself. Thus, if one hymn invokes Inanna/Venus as the "Inana, great light, lioness of heaven,"[325] another early hymn—*Inanna and Ebih*—credits the lioness with wreaking much destruction and slaughter:

"Goddess of the fearsome divine powers, clad in terror...drenched in blood, rushing around in great battles...covered in storm and flood, great lady Inana...In heaven and on earth you roar like a lion and devastate the people."[326]

Here it is significant to note that the Sumerian signs translated above as "great light"—ud-gal—are elsewhere used to signify "hurricane, or raging storm." Indeed, there is much reason to believe that the passage has been mistranslated and that the true meaning is "great storm" instead of "great light" (ud denotes "storm" as well as "sun" or sun-like brightness). Inanna herself is elsewhere invoked as "Great fierce storm," wherein storm is the aforementioned ud.[327] Yet another hymn describes Inanna as "Clothed (?) in a furious storm, a whirlwind."[328] Far from being the exception, such imagery is commonplace in Sumerian descriptions of the warring planet-goddess:

"Raining blazing fire down upon the land...At your battle cry, my lady, the foreign lands bow down. When humanity comes before you in awed silence at the terrifying radiance and tempest...Because of you, the threshold of tears is opened, and people walk along the path of the house of great lamentations. In the van of battle, all is struck down before you...You charge forward like a charging storm. You roar with the roaring storm."[329]

Now I ask: Would anyone viewing the planet Venus in its current orbit about the Sun ever be inspired to describe it in such fashion? Yet the fact remains that Venus was associated with destructive storms by other cultures far removed from Mesopotamia. Recall the Yakut description of Venus:

"It is said to be 'the daughter of the Devil and to have had a tail in the early days.' If it approaches the earth, it means destruction, storm, and

---

[324] J. V. Kinnier-Wilson, *The Rebel Lands* (Cambridge, 1979) represents a notable exception in this regard. He would understand the catastrophic imagery surrounding Inanna/Ishtar as inspired by a spectacular eruption of natural gas.

[325] Line 2 from "A hymn to Inana as Ninegala (Inana D)," *ETCSL*.

[326] Lines 1-7 from "Inana and Ebih," *ETCSL*.

[327] Line 1 from "A *balbale* to Inana (Inana A)," *ETCSL*.

[328] Line 21 from "A hymn to Inana (Inana C)," *ETCSL*.

[329] Lines 13-29 from "The exaltation of Inana (Inana B)," *ETCSL*.

frost, even in the summer; 'Saint Leontius, however, blessed her and thus her tail disappeared.'"[330]

It is also relevant to note that the mourning goddess's hair is specifically linked to a whirling storm. Witness the aforementioned passage from *Dumuzi's Dream* describing Geshinanna's lamentations for Dumuzi, quoted here in full: "My hair will whirl around in heaven for you like a hurricane."[331] Such traditions suggest that the disheveled "hair" of the mourning goddess was a visible and terrifying celestial phenomenon, something akin to the whirling tail of a great tornado.

Kali

Ancient India offers several notable examples of the warrior-goddess, the most famous of which is Kali. Renowned for her appetite for destruction and protruding tongue, the following description of the rampaging goddess is representative:

"Her anger grew so terrible that she transformed herself, grew smaller and black and left her lion mount and starting walking on foot. Her name then became Kali. With tongue lolling and dripping with blood, she then went on a blind destructive rampage, killing everything and everyone in sight, regardless of who they were."[332]

Although Kali is occasionally represented as beautiful, it is more common to find her depicted as repulsive in appearance. David Kinsley offered the following portrait of the Indian goddess:

"Hindu texts referring to the goddess are nearly unanimous in describing her as terrible in appearance and as offensive and destructive in her habits. Her hair is disheveled, her eyes red and fierce, she has fangs and a long lolling tongue, her lips are often smeared with blood, her breasts are long and pendulous, her stomach is sunken, and her figure is generally gaunt. She is naked but for several characteristic ornaments: a necklace of skulls or freshly cut heads, a girdle of severed arms, and infant corpses as earrings."[333]

As battle was deemed to be the "dance" of Inanna so too does Kali dance while waging war:

---

[330] L. Mándoki, *op. cit.*, p. 489.
[331] Translation in B. Alster, "The Mythology of Mourning," *Acta Sumerologica* 5 (1983), p. 6.
[332] J. Kripal, "Kali's Tongue and Ramakrishna," *History of Religions* 34:2 (1994), p. 161.
[333] D. Kinsley, "Blood and Death Out of Place: Reflections on the Goddess Kali," in J. Hawley and D. Wulff, eds. *The Divine Consort* (Berkeley, 1982), pp. 144-145.

> *"Ever art you dancing in battle, Mother. Never was beauty like thine, as with thy hair flowing about thee, thou dost ever dance, a naked warrior on the breast of Shiva."*[334]

Confronted with ancient literary reports that the dance of Ishtar/Venus threatened to destroy the very foundations of the world, scholars have typically understood them as the product of poetic license and figurative language. Yet the very same dire consequences are said to accompany Kali's terrible dance:

> *"The dread mother dances naked in the battlefield, Her lolling tongue burns like a red flame of fire, Her dark tresses, fly in the sky, sweeping away sun and stars, Red streams of blood run from her cloud-black limbs, And the world trembles and cracks under her tread."*[335]

As evidenced by this quote, Kali's disheveled hair was explicitly linked to an apocalyptic disaster involving the disappearance of the Sun. According to Alf Hiltebeitel, the goddess's "disheveled hair is thus itself an image of *Kalaratri*, the Night of Time, the night of the dissolution (*pralaya*) of the universe."[336]

There is a recurring emphasis in the Hindu texts on the disheveled hair of Kali. Indeed, an epithet of the warrior goddess—*Muktakesi*—commemorates her loosened and disheveled hair.[337] When it is reported that Kali's "streaming tresses hang in vast disorder,"[338] or that her disheveled hair blackens the skies, "sweeping away sun and stars," is it not evident that we are once again confronted by the mythology attached to comets?

As grotesque as Kali appears to the Western reader, her cult continues to exert a strange fascination over the people of India. Thus, Heinrich Zimmer described her as "today the most cherished and widespread of the personalizations of Indian cult."[339]

Bizarre as it is, Kali's monstrous form has striking parallels throughout the ancient world. Consider the ghoulish specter presented by the Aztec mourning goddess Itzpapalotl who, like Inanna, was conceptualized as a raging warrior:

> *"Obsidian Knife Butterfly [Itzpapalotl] is a wholly Chichimec goddess and her only office was war. She is depicted with a defleshed face and talons for feet and hands; she is winged and is often shown sweeping*

---

[334] D. Kinsley, *op. cit.*, p. 144.
[335] R. Tagore, *Sacrifice and Other Plays* (Bombay, 1917), p. 109.
[336] A. Hiltebeitel, "Draupadi's Hair," in M. Biardeau ed., *Autour de la déesse Hindoue* (Paris, 1981), p. 207.
[337] See J. Dowson, *A Classical Dictionary of Hindu Mythology and Religion* (London, 1961), p. 87.
[338] D. Kinsley, *The Sword and the Flute* (Berkeley, 1975), p. 120.
[339] H. Zimmer, *Myths and Symbols in Indian Art and Civilization* (Princeton, 1972), p. 215.

down from the heavens like a ghastly tzitzimitl. We are not shocked to see her in this form, but it comes as something of a shock to see her also cast in mythology as a double of Precious Flower [i.e., Xochiquetzal, the Aztec Aphrodite]...This is an outstanding example of the interpenetrability of the forms of the Great Mother."[340]

The comparison of the mourning Itzpapalotl to a Tzitzimitl is most interesting for, as Brundage notes, the latter creature "is an eerie goddess in the night sky...[whose] hair is madly disheveled."[341] A picture of a Tzitzimitl is depicted in figure 1.[342] Notice the grotesquely protruding tongue and necklace of hearts and hands. The uncanny resemblance to Kali is readily apparent.

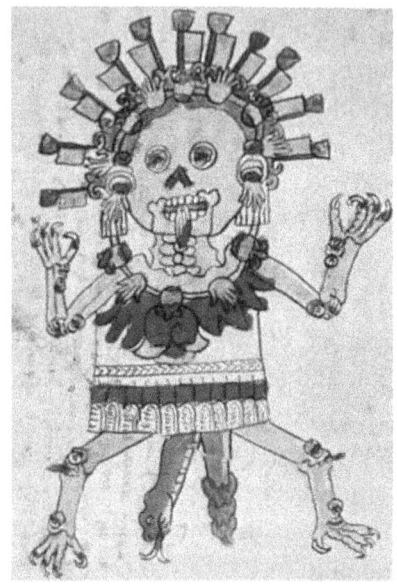

Figure 1

According to Aztec tradition, Itzpapalotl was hurled from heaven for sinning against the gods.[343] This tradition finds a remarkable parallel in ancient Mesopotamia, where it was reported that Lamashtu—a recognized avatar of Inanna/Ishtar[344]—was hurled from heaven for her evil deeds. It was on that occasion that the demonic goddess displayed wildly disheveled hair. An Assyrian incantation alludes to this event:

---

[340] B. Brundage, *The Fifth Sun: Aztec Gods, Aztec World* (Austin, 1983), p. 173.
[341] *Ibid.*, p. 62.
[342] Adapted from the *Codex Magliabecchiano*.
[343] B. Brundage, *op. cit.*, p. 46, citing the *Codex Vaticanus* plate 43.
[344] W. Fauth, "Ištar als Löwingöttin und die löwenköpfige Lamaštu," *Die Welt des Orients* 12 (1981), pp. 33-34.

"She is a haunt, she is malicious, Offspring of a god, daughter of Anu. For her malevolent will, her base counsel, Anu her father dashed her down from heaven to earth, For her malevolent will, her inflammatory counsel. Her hair is askew, her loincloth is torn away."[345]

The image of Ishtar-Lamashtu being hurled from heaven with disheveled hair once again evokes cometary imagery.[346] Lamashtu's disheveled hair and tattered clothes, likewise, recall the appearance traditionally accorded mourners.

A witch-like goddess renowned for her chimeric form and ogre-like appetites, Lamashtu was said to have the head of a lion:

"Great is the daughter of Anu...She is cruel, raging, wrathful, rapacious...Her head is the head of a lion."[347]

Yet another hymn describes the goddess as having a leonine face and disheveled hair:

"She is furious, she is fierce, she is uncanny, she has an awful glamor... the daughter of Anu!...The face of a ravening lion is her face. She came up from the reed bed, her hair askew..."[348]

Over the course of untold centuries, Lamashtu eventually became demonized to a point at which her original identification with Inanna/Ishtar was all but forgotten. The transformation of the great mother goddess into a witch was complete:

"Among all the devils and fiends of which the Mesopotamians lived in terror, the one that seems to have been the most dreaded was [Lamashtu], a she-devil, and the daughter of the great god Anu...The goddess Lamashtu was a violent, raging devil of terrifying aspect... With her hair tossed about wildly, and her breasts uncovered she burst out of the cane brakes like a whirlwind..."[349]

The fact that the raging warrior-goddess with disheveled hair can be found in the New World as well as the Old is compelling *prima facie* evidence that the imagery in question originated as a direct result of common experience—most likely a particularly memorable comet-like apparition. Yet as the example provided by Ishtar-Lamashtu attests, there is also an indissoluble connection with the planet Venus. Here, too, New World traditions provide a corroborating

---

[345] B. Foster, *Before the Muses: An Anthology of Akkadian Literature* (Bethesda, 1993), p. 59.

[346] As Carl Sagan remarked in *Comet* (New York, 1985), p. 14: "A comet suggests flowing tresses." See also the discussion in W. Gundel, "Kometen," *RE*, col. 1175-1176.

[347] B. Foster, *op. cit.*, p. 865.

[348] *Ibid.*, p. 864.

[349] E. Budge, *Amulets and Talismans* (New York, 1968), pp. 104-109.

parallel. Thus, an Inca name for Venus was *chasca coyllur*, signifying the "star (*coyllur*) with tangled or disheveled hair."[350] The modern descendants of the Inca, moreover, continue to observe "the day of disheveled hair," presumably because of its cosmological import: "In the Andes, the modern lexicographer Lara has noted a Quechua neologism, *ch'askachau*—literally 'the day of disheveled hair'—meaning *viernes*, the Spanish word for Venus's day."[351]

The conclusion seems inescapable: It was the planet Venus itself, conceptualized as a raging warrior-goddess, which once displayed disheveled "hair" while participating in a spectacular cataclysm that shook the foundations of heaven and earth and blotted out the light of the Sun. Whether memorialized as the lurid lamentations of Inanna, the battle dance of Ishtar, or Kali's apocalyptic "night of the dissolution of the universe," the various mythological traditions describe a raging planetary-goddess hellbent on destruction.

---

[350] W. Sullivan, *The Secret of the Incas* (New York, 1996), p. 87, citing Diego Holguin's *Vocabulario de la lengua general de todo el Peru llamada lengua Quichua o del Inca*.
[351] *Ibid.*, p. 88.

10. The "Witch-Star"

"The ultimate origin of nearly all folktales and myths must remain a mystery, just as the origin of language is a mystery."[352]

With the retreat of Old European religions under the onslaught of Christianity, the heathen goddesses suffered a profound diminution in their status and sphere of influence. Not to be forgotten or eradicated entirely, they eventually resurfaced in modified form as demons or as leading characters in folklore and fairy tales. The curious traditions surrounding witches are a case in point. For untold centuries, in Old Europe as around the globe, witches have been credited with the ability to fly through the sky, raise storms, cast spells, poison, and blot out the sun. As Jacob Grimm documented, the folklore associated with witches is of great antiquity and reflects fundamental aspects of the cult of the mother goddess: "The details of witchcraft, the heart-eating, the storm-raising, the riding through air, are all founded on very ancient and widely scattered traditions."[353]

It is understandable that the modern reader might well be skeptical of discovering much that is of historical or scientific import in ancient and medieval traditions surrounding witches and the instruments of witchcraft. After all, the skies today are not blackened by the spectacle of disheveled hags flying about, as in *The Wizard of Oz*; and the days when the mass of humanity believed that witches controlled the weather or caused eclipses

---

[352] S. Thompson, "Myth and Folktales," in T. Sebeok ed., *Myth: A Symposium* (Baltimore, 1955), p. 176.
[353] J. Grimm, *Teutonic Mythology, Vol. 3* (Gloucester, 1976), p. 1077.

of the Sun are long behind us. It might also appear as if we have strayed far afield from our primary subject matter—namely, ancient conceptions surrounding the planet Venus.

Yet it is a demonstrable fact that many of the world's great goddesses are frequently described in terms otherwise befitting a witch. Ishtar-Lamashtu, as we have seen, was represented as a hag-like demon with disheveled hair, swooping down from the sky and making off with the neighborhood children. The same archetypal figure, under the name of Lilith, played a significant role in ancient and medieval Jewish lore, wherein she was represented as a child-stealing witch with horribly disheveled hair.[354] A magical text from the first millennium A.D. contains the following curse directed at Lilith: "Naked shall you be driven away, unclothed, with your hair loose and streaming behind your back."[355]

The Norse Freyja, in addition to her avatars as warrior and lamenting goddess, also appears as a witch. In the *Völuspa*, for example, the goddess (as Gullveig) is invoked as follows: "Witch was her name in the halls that knew her, a sorceress, casting evil spells."[356]

The great goddess assumes the form of a witch in Slavic lore as well. Marija Gimbutas offered a representative example of this mythical archetype from Russia:

> *"Baba Yaga, the ancient Goddess of Death and Regeneration in Slavic mythology, is well preserved in folk tales (mainly Russian) in a degraded form, i.e., as a witch. She might be depicted as an evil old hag who eats humans, especially children, or as a wise, prophetic old woman. In appearance, she is tall, bony legged, pestle headed, and has a long nose and disheveled hair."*[357]

Although witch-like characteristics can be found within the cults of most great goddesses, it is the Germanic Holda and Greek Hecate who offer the most revealing portraits of the Witch-Goddess.

Holda

Amongst the various Germanic tribes, Holda/Holle is remembered as a disheveled crone flying through the night-sky and dispensing misfortune. While

---

[354] S. Hurwitz, *Lilith—The First Eve* (Einsiedeln, 1992), p. 103, describes Lilith as follows: "She is mostly depicted as naked, with prominent breasts and unbound hair that streams wildly behind her back."

[355] *Ibid.*, p. 96.

[356] Volüspa 22 as translated by P. Terry, *Poems of the Elder Edda* (Philadelphia, 1990), p. 2.

[357] M. Gimbutas, *The Language of the Goddess* (San Francisco, 1989), p. 210.

riding on the wind, she is said to be "clothed in terror" and accompanied by a "furious host," the latter composed of disembodied souls and other ghoulish beings.[358] Holle-riding, "to ride with Holle," was deemed equivalent to the nocturnal ride of witches.[359]

Despite a bloody, centuries-long crusade against heathen religion and witchcraft, the agents of the Christian church never did succeed in wholly eradicating the memory of the Germanic goddess. Rather, Dame Holda lived on in popular consciousness, particularly in German fairy tales. Gimbutas summarized the situation as follows:

"In spite of the horrible war against women and their lore and the demonization of the Goddess, the memories of her live on in fairy tales, rituals, customs, and in language. Collections such as Grimm's German tales are rich in prehistoric motifs describing the function of this Winter Goddess Frau Holla (Holle, Hell, Holda, Perchta, etc.). She is the ugly Old Hag with a long nose, large teeth, and disheveled hair."[360]

Holda is alternately described as both beautiful and as an ugly hag. According to a fable cited by Jacob Grimm, the goddess's transformation into a witch occurs in the blink of an eye:

"Hulda, instead of her divine shape, assumes the appearance of an ugly old woman, long-nosed, big toothed, with bristling and thick-matted hair. 'He's had a jaunt with Holle', they say of a man whose hair sticks up in tangled disorder; so children are frightened with her or her equally hideous train."[361]

Disheveled hair, it will be noted, is a prominent feature of the Witch-Goddess. Indeed, the epithet *Werra* is thought to commemorate Holda's "tangled, shaggy hair."[362]

Although later legends humanized Holda, envisioning her as a spinning wife, enchantress, or spirit of the local water-spring, Grimm points out that the Germanic goddess was originally located in the sky.[363] Significantly, Grimm places the identification of Holda/Holle with the Latin Venus as "beyond question."[364]

In Norwegian and Danish folk-tales, Holda was known as Huldra. Like Holda, the Danish demon was described as a witch with a penchant for making

---

[358] J. Grimm, *Teutonic Mythology*, Vol. 1 (Gloucester, 1976), pp. 265-272.
[359] *Ibid.*, p. 269.
[360] M. Gimbutas, *op. cit.*, p. 319.
[361] J. Grimm, *op. cit.*, p. 269.
[362] *Ibid.*, p. 273.
[363] *Ibid.*, p. 267.
[364] *Ibid.*, Vol. 3, p. 935.

off with children. Otherwise beautiful, Huldra had a most peculiar physical feature—a tail!

Hecate

The patron-goddess of witches and sorceresses, Hecate was one of the most popular goddesses in all of Greece. First attested in Hesiod, where she is described as an all-powerful mother goddess, Hecate is equally at home in heaven or on earth.[365] Hecate was invoked as a goddess of childbirth and marriage, and early epithets confirm that she had a benevolent aspect (*Kourotrophos*, among others). Other epithets mark her as a goddess of the crossroads (*Trivia*) and protectress of doors and entrances (*Epipyrgidia*).[366]

In a development analogous to that which saw Inanna/Ishtar transformed into the witch-like Lamashtu, Hecate eventually came to be regarded as the epitome of the terrible goddess—"the principal source and originator of all that was ghostly and uncanny."[367] The Witch-Goddess was said to fly about at night on the wind while brandishing torches.[368] Like Holda, Hecate was intimately associated with a train of souls and howling dogs, the latter said to surround the goddess on her nocturnal jaunts: "Queen of the spirits of the dead, she was active at night, accompanied by a retinue of dogs and ghosts of suicides or those who had died a violent death."[369]

Hecate was elsewhere described as a cannibal. An excerpt from the *Greek Magical Papyrus* emphasizes her demonic nature:

"*O nether and nocturnal, and infernal Goddess of the dark...O you with hair of serpents, serpent-girded, who drink blood, who bring death and destruction, and who feast on hearts...*"[370]

The cannibalism ascribed to Hecate recalls the ogreish behavior attributed to Kali, Anat, Baba Yaga, and the Samoan Venus (*Tapuitea*). Yet H. J. Rose, together with other scholars, called attention to Hecate's fundamental affinity with Aphrodite.[371] This is baffling at first sight, for what could the Witch-Goddess have to do with the Queen of Heaven? By now the reader can guess the

---

[365] *Theogony* 411-452.
[366] Aristophanes, *Vesp.* 804. See also the discussion in C. Faraone, *Talismans and Trojan Horses* (New York, 1992), pp. 8-9.
[367] E. Rhode, *Psyche, Vol. 2* (New York, 1966), p. 297. Rohde's summary remains the most insightful analysis of Hecate's cult to this day.
[368] L. Farnell, *The Cults of the Greek States, Vol. 2* (New Rochelle, 1977), p. 505.
[369] V. Newall, *The Encyclopedia of Witchcraft and Magic* (New York, 1974), p. 94.
[370] 4:2854-67 as translated in J. Rabinowitz, *The Rotting Goddess* (Brooklyn, 1998), p. 62.
[371] H. Rose, *A Handbook of Greek Mythology* (New York, 1959), p. 122.

answer: The Witch, like the Queen of Heaven, has her origin in the prehistoric appearance of the planet Venus.

Nahtfare

If indeed the witch-like characteristics ascribed to the great goddesses reflect ancient conceptions associated with the planet Venus, one would not be surprised to find an explicit connection between that planet and witchcraft. On this score the ancient sources do not disappoint. Thus it is that Babylonian astronomical texts denote the planet Venus as the "witch-star" (*kakkab kaššāptu*).[372] Astronomers will be hard-pressed to find a rational explanation for this bizarre epithet applied to Venus apart from the thesis offered here.

A reminiscence of the planet Venus as witch is also apparent in ancient Norse lore, as Grimm pointed out long ago: "There is perhaps more of a mythic meaning in the name *nahtfare* for evening star (Heumanni opusc. 453. 460), as the same word is used of the witch or wise-woman out on her midnight jaunt."[373]

Arabic lore likewise portrays Venus as a "witch-star." According to the Islamic *Book of Idols*, Mohammed once entered the sacred precinct of al-'Uzza and chopped down a tree beloved by the planet-goddess, thereby arousing her wrath. As a result of this sacrilege, al-'Uzza assumed the form of a witch with wildly disheveled hair and gnashing teeth.[374]

Students of ancient myth and folklore have long struggled to make sense of the complex imagery surrounding witches by reference to the natural world. As a last resort, attempts have been made to find an explanation for the "terrible goddess" in terms of subjective psychological factors. Erich Neumann's analysis is typical in this regard:

"The symbolism of the Terrible Mother draws its images predominantly from the 'inside'; that is to say, the negative elementary character of the Feminine expresses itself in fantastic and chimerical images that do not originate in the outside world. The reason for this is that the Terrible Female is a symbol for the unconscious. And the dark side of the Terrible Mother takes the form of monsters, whether in Egypt or India, Mexico or Etruria, Bali or Rome. In the myths and tales of all peoples, ages, and countries—and even in the nightmares of our own nights—witches and vampires, ghouls and specters, assail us, all terrifyingly alike."[375]

---

[372] F. Gössmann, *Planetarium Babylonicum* (Rome, 1950), p. 62.
[373] J. Grimm, *Teutonic Mythology*, Vol. 2 (Gloucester, 1976), p. 723.
[374] M. Höfner, "Die vorislamischen Religionen Arabiens," in H. Gese et al eds., *Die Religionen Altsyriens, Altarabiens und der Mandäer* (Stuttgart, 1970), p. 363.
[375] E. Neumann, *The Great Mother* (Princeton, 1974), pp. 148-149.

Our hypothesis would turn that of Neumann on its head: the archetypal images of the terrible goddess—Lamashtu, Kali, Holda, Hecate, the witch—have an objective basis in the real world, tracing ultimately to the prehistoric appearance of the planet Venus while displaying a comet-like appearance. In the archaic tradition of Inanna/Lamashtu falling from heaven with disheveled hair—understood here as a memorable phase in the natural history of the planet Venus—we recognize the celestial prototype for the witch as portrayed in ancient myth and folklore. The widespread beliefs surrounding witches are "all terrifyingly alike"—to use Neumann's phrase—precisely because they originated in the "outside world" and were witnessed by cultures the world over. Witches were distinguished by their disheveled hair precisely because Venus itself displayed "disheveled" hair. And as witches were believed to be capable of raising storms by unbinding their hair so, too, was Venus described as an agent of storm and storm-raiser extraordinaire.[376] The extraordinary powers credited to witches, such as the ability to "fly" through the sky or effect an eclipse of the Sun, are best understood as vestigial echoes of ancient traditions describing the terrifying behavior associated with the Venus-comet. From the standpoint of comparative religion, therefore, the witch is simply a condensation or watered-down version of the warring or mourning goddess, roaming the world with disheveled hair while wreaking destruction and bringing storms, threatening to blot out the Sun forever.

---

[376] B. Walker, *The Woman's Encyclopedia of Myths and Secrets* (San Francisco, 1983), p. 368 notes the "unflagging superstitious belief in Christian Europe that witches' hair controlled the weather. Churchmen said witches raised storms, summoned demons, and produced all sorts of destruction by unbinding their hair. As late as the 17th century the *Compendium Maleficarum* said witches could control rain, hail, wind, and lightning in such a way."

11. Sovereignty as Hag:
A Case Study in Mythological Analysis

> *"Stories of a king's, or a potential king's, lovemaking with the goddess of Sovereignty are so widespread in early Ireland and elsewhere in Europe, such as Geoffrey Chaucer's 'Wife of Bath's Tale', as to merit their own international folk-motif number, D732. According to the conventionalized steps in the story, the male protagonist encounters an ugly hag who invites him to have intimate relations with her. Her repulsiveness...initially put him off, but he eventually relents. On the morning after their lovemaking, the hag is transformed into a beautiful maiden."*[377]
>
> *"When one inquires what kind of stories are these which have been credited with such extraordinary power, one finds that they tell of the adventures of heroes and heroines; enchantments and disenchantments; kings and queens, ogres and monsters and fairies...Such is the stuff of the stories told by all peoples whose traditional culture has not been upset by the teaching of modern history and modern science, and it is remarkable how the same themes or motifs, and even series of motifs, recur in the traditions of peoples widely separated from one another in space and in time. The very homogeneity of the material presents a considerable problem to the modernist. What is there in this fantastic heritage that from time immemorial it should have retained the sympathy and excited the wonder of mankind?"*[378]

A popular theme in Celtic lore finds Sovereignty being personified as a haggish woman.[379] According to the *Táin Bó Cuailnge* and a number of other medieval texts, a would-be king can only obtain the throne by kissing or "marrying" a hideous-looking hag who, in various versions of the tale, suddenly transforms into a beautiful woman shortly thereafter.[380] Although the

---

[377] J. MacKillop, *Dictionary of Celtic Mythology* (Oxford, 1998), p. 344.

[378] A. Rees & B. Rees, *Celtic Heritage* (London, 1961), pp. 19-20.

[379] The theme was so popular, in fact, that James Joyce made it a central motif in his *Ulysses*. See M. Tymoczko, *The Irish Ulysses* (Berkeley, 1994), pp. 106-119.

[380] See the discussion in A. Coomaraswamy, "On the Loathly Bride," in R. Lipsey ed., *Coomaraswamy: Selected Essays* (Princeton, 1977), pp. 353-370; L. Sumner, "The Weddynge of Sir Gawen and Dame Ragnall," in *Smith College Studies in Modern Language* V:4

oldest extant version of the legend evidently dates from the eleventh century, scholars are generally agreed that the thematic pattern is archaic in nature.[381]

In the *Book of Ballymote*, a manuscript dating from the 14th century, it is recounted how one king Daire, the George Foreman of his time, named each of his five sons Lugaid since it had been prophesied that a hero of this name would one day rule Ireland. In the tale in question, the last born son, Lugaid Laigde, journeyed to a distant house wherein there dwelled "a huge old woman...her spears of teeth outside her head, and great, old, foul, faded things upon her."[382] Upon joining her in bed, the hero was witness to a dramatic metamorphosis in the hag's appearance and demeanor:

"Lugaid Laigde who, with astonishment, saw the old body, under his embrace, become radiant like the rising sun in the month of May and fragrant like a beautiful garden. As he clasped her, she said to him: 'Happy is your journey, for I am Sovereignty, and you shall attain the sovereignty over all of Ireland.'"[383]

Similar stories were told across the European continent. The hag's horrible appearance is a recurring point of emphasis, with her dark form and disheveled hair standing out.[384] According to *The Adventure of the Sons of Eochaid Muigmedón*, the hag was discovered guarding a well:

"Thus was the hag: every joint and limb of her, from the top of her head to the earth, was as black as coal. Like the tail of a wild horse was the grey bristly mane that came through the upper part of her head...Dark, smoky eyes she had, and a nose crooked and cavernous. Her middle was fibrous, spotted with pustules, and diseased...Loathsome, indeed, was the hag's appearance."[385]

Yet once the hero Niall kisses the hideous creature she immediately turns into the most beautiful woman in the world. At that point she announces: "I am Sovereignty."

(1924), pp. vii-39; T. O'Maille, "Medb," *Zeitschrift für celtische Philologie* 17 (1958), pp. 129-146; A. Krappe, "The Sovereignty of Erin," *The American Journal of Philology* 63:4 (1942), pp. 444-454; A. Brown, *The Origin of the Grail Legend* (Cambridge, 1943), pp. 210-224.

[381] G. Dumezil, *The Destiny of the King* (Chicago, 1973), p. 98.

[382] Quoted from A. Hiltebeitel, *The Ritual of Battle* (Ithaca, 1976), pp. 175-176.

[383] Quoted from G. Dumezil, *op. cit.*, pp. 89-90.

[384] A. Eichhorn-Mulligan, "The Anatomy of Power and the Miracle of Kingship," *Speculum* 81:4 (2006), p. 1031 observed that the hag was characterized by her "dark, blackened appearance; and unruly hair."

[385] As translated by W. Stokes and reprinted in A. Bourke et al eds., *The Field Day Anthology of Irish Writing, Vol. IV* (New York, 2002), p. 261.

A number of hypotheses have been advanced in an attempt to explain the origins of this peculiar constellation of ideas. A connection with the archaic rite of the *hieros gamos* is commonly acknowledged.[386] According to T. F. O'Rahilly, the myth survives as a vestigial memory of archaic mystery rites involving the king's marriage to Mother Earth herself:

> *"The idea that Ireland is a goddess, and is wedded to the king of the country, is of hoary antiquity…It has its roots in the time when men regarded the material Earth as a Mother, and when the ruler of the land was inaugurated with a ceremony which professed to espouse him to this divine mother, with the intent that his reign be prosperous and that the earth might produce her fruits in abundance."*[387]

For Ananda Coomaraswamy, the various legends presenting Sovereignty as a Hag have a wholly mundane explanation. According to this esteemed scholar, the myth of the loathly bride has its origin in the perpetual war between the respective sexes: "We have so far seen that the heroic motif of the transformation of a hideous and uncanny bride into a beautiful woman cannot be regarded as peculiarly Celtic, but rather represents a universal mythical pattern, underlying all marriage, and one that is, in fact, the 'mystery' of marriage."[388]

In an insight-laden commentary on the loathly bride myth, Coomaraswamy made much of the fact that, in several variants of the tale, the hag is given serpentine features or otherwise likened to a snake: "The Loathly Lady must be identified with the Dragon or Snake whom the hero disenchants by the Fier Baiser."[389] Defending his beloved field of metaphysics against more profane explanations of ancient myth, Coomaraswamy cautioned against any and all attempts to refer such universal themes to a historical prototype:

> *"Myths are significant, it will be conceded: but of what? If we do not ask the right questions, with the Grail before our eyes, our experience of the mythical material will be as ineffectual as that of the hero who reaches the Grail castle and fails to speak, or that of the hero who will not kiss the Dragon: our science will amount to nothing more than the accumulation of data, which can be classified, but cannot be brought to life. Myths are not distorted records of historical events. They are not periphrastic descriptions of natural phenomena, or 'explanations' of them, so far from that, events are demonstrations of the myths."*[390]

---

[386] A. Eichhorn-Mulligan, *op. cit.*, p. 1054.
[387] T. F. O'Rahilly, "On the Origin of the Names *Erainn* and *Eriu*," *Eriu* XIV (1943), p. 21 as quoted in S. Eisner, *A Tale of Wonder* (London, 1957), p. 17.
[388] A. Coomaraswamy, *op. cit.*, p. 363.
[389] *Ibid.*, p. 355.
[390] *Ibid.*, p. 368.

Alas, Coomaraswamy never got around to explaining why the loathly hag came to be conceptualized as a serpentine dragon, much less why Sovereignty itself should be envisaged as a woman or as something that could be won through marriage. Other equally perplexing questions remain unexplained to this very day. Whence derives the hag's disheveled hair and repellent form in the first place? And how we are to explain the hag's sudden transformation into a beautiful woman?

It is our opinion, in contrast to Coomaraswamy, that the serpentine form of the loathly hag—herself the very embodiment of sovereignty—will never be explained by reference to the interpersonal relationships of men and women. Nor, for that matter, is it preferable to refer the peculiar imagery surrounding the hag to unconscious determinants as per Neumann or Jung as that merely replaces one mystery with another and begs the following question: Whence derives the specific contents or mythological imagery to be found in the unconscious, collective or otherwise?

The discerning reader has doubtless deduced the correct answer: It is the ancient conceptions associated with the planet Venus that point the way to a successful analysis. Those readers who are familiar with my writings will remember that Venus was identified as the source of sovereignty by mythmakers around the globe. In the early Sumerian epic poem *Enmerkar and the lord of Aratta*, the legendary king Enmerkar is made to announce: "The ever-sparkling lady gives me my kingship."[391] The word translated as "ever-sparkling" here is mul-mul-e, "to radiate, or shine," a verb formed from the Sumerian word for star (mul) and hence referring to the brilliant splendor of Venus. The clear import of this passage, accordingly, is that it is the planet Venus which invests the stellar hero Enmerkar with kingship.

The same basic idea is found in other Sumerian texts as well. According to *A Song of Inana and Dumuzi*, it is the embrace of Inanna provides the aspiring king with sovereignty:

"May the lord whom you have chosen in your heart, the king, your beloved husband, enjoy long days in your holy and sweet embrace! Give him a propitious and famous reign, give him a royal throne of kingship on its firm foundation, give him the scepter to guide the Land, and the staff and crook, and give him the righteous headdress and the crown which glorifies his head!"[392]

Inanna's function as the goddess of Sovereignty is also a point of emphasis in the so-called sacred marriage rite. According to Iddin-Dagan's hymn describing

---

[391] Line 632 from "Enmerkar and the lord of Aratta," *ETCSL*.
[392] Lines 34-40 in "A song of Inana and Dumuzi (Dumuzid-Inana D1), *ETCSL*.

the rite in question, it was commonly believed that the king could only obtain the throne by "marrying" or engaging in simulated sexual intercourse with the planet Venus. That the "marriage" itself involves a conjunction of celestial powers is indicated by the fact that the brilliantly shining Inanna/Venus is said to embrace a "sun-like" king:

> *"After the lady has made him rejoice with her holy thighs on the bed, after holy Inana has made him rejoice with her holy thighs on the bed, she relaxes (?) with him on her bed... She embraces her beloved spouse, holy Inana embraces him. She shines like daylight on the great throne dais and makes the king position himself next(?) to her like the sun."*[393]

Such imagery naturally recalls the Celtic tale of Lugaid's encounter with the Hag of Sovereignty, whose "embrace" makes him king. Upon mating with the hero, the Hag is suddenly transformed into a radiant celestial form:

> *"Lugaid Laigde who, with astonishment, saw the old body, under his embrace, become radiant like the rising sun in the month of May and fragrant like a beautiful garden."*[394]

The comparison of the Hag to a "beautiful garden," in turn, recalls the fact that Inanna/Venus herself was repeatedly compared to a luxuriant garden in Sumerian hymns recounting the sacred marriage rite.[395] Indeed, it is reported that the consummation of Inanna's marriage with Dumuzi was accompanied by the sudden appearance of a verdant garden:

> *"The holy embrace...As she arises from the king's embrace the flax rises up with her, the barley rises up with her. With her, the desert is filled with a glorious garden'."*[396]

As I have documented elsewhere, analogous traditions will be found around the globe, in the New World as well as the Old. In ancient Persia, for example, it was the goddess Anahita—expressly identified with the planet Venus—that provided the king with the crown of sovereignty: "Anahita appears as a granter of sovereignty and glory (Aban Yast, XII, 46; XIII, 50), and is portrayed in various media handling investiture crowns to kings."[397]

In ancient India it was the goddess Śri who embodied sovereignty. As was the case with Inanna/Venus, early kings were believed to enter into conjugal

---

[393] Lines 193-202 as translated in J. Black et al, *The Literature of Ancient Sumer* (Oxford, 2004), p. 267.
[394] Quoted from G. Dumezil, *op. cit.*, pp. 89-90.
[395] Y. Sefati, *Love Songs in Sumerian Literature* (Jerusalem, 1998), p. 34. See also P. Lapinkivi, *The Sumerian Sacred Marriage Rite* (Helsinki, 2004), p. 39.
[396] Lines C 1-11 from "A *balbale* (?) to Inana (Dumuzid-Inana P)," *ETCSL*.
[397] Y. Ustinova, "Aphrodite Urania," *Kernos* 11 (1998), p. 218.

relations with the goddess in order to obtain the throne.[398] Most significant, perhaps, is the fact that, as the goddess of sovereignty, Śrī was thought to be incarnate in the royal diadem, thereby paralleling the traditions surrounding Inanna.[399] Of this goddess, Coomaraswamy remarked:

> *"Śrī ('Splendor')-Lakshmi ('Insigne') is the well-known Indian Goddess of Fortune (Tyche), Prosperity (the personified 'Luck" of western folklore) and Beauty: she is the principle and source of all nourishment, kingship, empire, royalty, strength, sacerdotal luster, dominion, wealth, and species, which are appropriated from her by the gods whose distinctive properties they are."*[400]

Despite her many positive attributes, Śrī had at least one unsettling trait: She was said to be capable of taking on the form of a serpent.[401]

Granted that the Hag's frightful appearance and serpentine form finds a close parallel in the archaic mythology and descriptors attached to Inanna/Venus and analogous goddesses like the uraeus-goddess and Kali, it remains to ask how we are to understand her sudden transformation into a beautiful woman upon marrying the lucky king-to-be? An important clue comes from ancient Egypt. In Egyptian texts, it is expressly stated that the goddess's "raging" threatened to destroy the world, much as was the case with the destructive rampage of Inanna/Venus. This archaic story is most familiar from the account in *The Destruction of Mankind*, which finds Hathor being dispatched by Re to rain fire and destruction upon mankind:

> *"Then mankind plotted something in the (very) presence of Re...Then they [Re's advisors] said in the presence of his majesty: 'May thy Eye be sent, that it may catch for thee them who scheme with evil things... It should go down as Hathor.' So then this goddess came and slew mankind in the desert."*[402]

After a period of indiscriminate slaughter, the Egyptian Eye goddess is magically "calmed" and returned to her consort—typically Horus or Ra—whom she then marries amidst much pomp and revelry. Most instructive for our purposes here is the report that the Eye-goddess assumed a beautiful form upon consummating her marriage:

---

[398] J. Gonda, *Aspects of Early Visnuism* (Delhi, 1969), p. 220.
[399] A. Hiltebeitel, "Draupadi's Hair," *Purasārtha* 5 (1981), pp. 191-192.
[400] A. Coomaraswamy, *op. cit.*, p. 358.
[401] J. Gonda, *op. cit.*, p. 221.
[402] J. Wilson, "Deliverance of Mankind from Destruction," in J. Pritchard ed., *The Ancient Near East* (Princeton, 1958), p. 4.

> *"Finally persuaded to return, the goddess arrives in a festival procession at Philae, where she purifies herself in the sacred waters of the Abaton, transforming into a beautiful woman whom Ra welcomes into his arms. With her pacification, the order of the cosmos is also restored."*[403]

A very similar turn of events occurs in the mythological history of Inanna/Venus as well. Thus it is that one hymn prays "May the great gods calm your mood."[404] So, too, in the final lines of Iddin-Dagan's marriage hymn, the warrior-goddess is said to be calmed or "soothed" as a result of the sexual union with her consort (it will be noted that the word kuš$_2$, translated as "relaxes?" by Jeremy Black in the passage quoted above, also denotes "to soothe or calm" and is well known as a euphemism for sexual intercourse in Sumerian sources).[405] Such, in essence, is the story of the Hag as Sovereignty: She is beautified as a result of sexual union with the would-be king.

Having identified the Hag of Sovereignty with the planet Venus, it remains to discover the celestial identity of her paramour, the would-be king. In the Celtic tale quoted at the outset of this study, the male suitor in question is Lugaid Laigde, whom scholars have recognized as a humanized version of Lug, the greatest Celtic god of all.[406] Early on identified with the Latin god Mars, Lug was also represented as the Lord of the Underworld, where he was allegedly attended by the Goddess of Sovereignty.[407] Lug's identification with Lugaid, as MacKillop argued, is rendered near certain by the fact that the god embodies kingship and his wife Sovereignty in an Irish tale known as *Baile in Scail* ("The Phantom's Frenzy").[408] As the greatest warrior-hero in

---

[403] B. Richter, "On the Heels of the Wandering Goddess," in M. Dolinska &. H. Beinlich eds., *Ägyptologische Tempeltagung* (Wiesbaden, 2010), p. 156. See also H. Junker, *Der Auszug der Hathor-Tefnut aus Nubien* (Berlin, 1911), p. 6, who remarks: "Auf dem Abaton kühlt Schu ihre Glut, und sie reinigt ihre Glieder im Wasser der heiligen Insel. Da wandelt sich die Löwin in eine holde Frau mit leuchtenden Augen und frohem Angesicht, mit Locken und Brüsten, die Herrin der Frauen, glänzend in ihrer Schönheit, mit fürstlicher Gestalt."

[404] Line 259 from "A hymn to Inana (Inana C)," *ETCSL*.

[405] Y. Sefati, *op. cit.*, p. 39, states: "In the context of the sacred marriage texts, the expression 'to soothe the heart' (ša kuš-u) is a euphemism for sexual union." See also J. Halloran, *Sumerian Lexicon* (Los Angeles, 2006), p. 153.

[406] A. Brown, *The Origin of the Grail Legend* (Cambridge, 1943), pp. 217-218 writes: "Eoin MacNeill has pointed out that most kings named Lughaid are merely faded pictures of the god, *Lug mac Eithne*. It is therefore probable that Lughaid Laighe, in the poem of *Carn Mail* above, has borrowed his adventure from the god Lug. We may even go further and infer that all the Irish stories just outlined probably began in stories about Lug. As Lug was the winner of sovereignty, it is natural that he should appear in some versions as lord of the great house where the sovereignty dwells."

[407] H. R. Davidson, *Myths and Symbols in Pagan Europe* (Syracuse, 1988), p. 89-91.

[408] J. MacKillop, *op. cit.*, p. 345.

Celtic lore and prototypical King *par excellence*, Lug is to be identified with the planet Mars.[409]

To summarize our findings: A satisfactory explanation of the Celtic legend of the Hag as Sovereignty continues to elude modern students of myth primarily because they have consistently overlooked its original celestial context. In the final analysis, the Celtic Hag is best interpreted as a vestigial survival of the witch or witch-star, the latter indistinguishable from the raging warrior-goddess (Inanna, the Egyptian Eye, Kali, etc.). The witch, in turn, represents the planet Venus during a period of spectacular instability, wherein it presented the appearance of a terrifying comet-like body with wildly disheveled hair—the prototypical star of *disaster*.

The metamorphosis of the Hag into the goddess of Sovereignty, on the other hand, traces to a subsequent phase in astronomical history during which Venus conjoined with or "married" Mars, as a result of which she was "calmed" and her terrifying form disappeared. It was during this post-coital phase that Venus's raging "hair" disappeared or was otherwise brought under control, thereby restoring order to the cosmos. With the "calming" of the raging planet-goddess and the binding or ordering of her unruly hair, the Queen of Heaven assumes a radiantly beautiful form as the crown or "glory" of the Sovereign King.

---

[409] E. Cochrane, *Martian Metamorphoses* (Ames, 1997), pp. 152-153.

12. Hathor: The Egyptian Venus

"The departure of the Eye, or Hathor, results in a series of natural disasters in Egypt, where perpetual night prevails."[410]
"The eye of the sun-god was an independent part of himself, with a complicated mythological history."[411]

The Egyptian Hathor offers an exemplary model of the mother goddess. Already in the third millennium BCE she was regarded as the mother of Horus, the Egyptian star-god believed to be incarnate in the pharaoh. Hathor's very name reflects this archaic relationship, signifying "House of Horus."[412]

Scholars have hitherto been at a loss to explain the original nature of this great goddess, puzzled not only by her name but by her numerous seemingly incompatible characteristics. Alison Roberts, in a recent study of Hathor, offered the following complaint: "My initial problem was how to find any coherent pattern in the many representations of the goddess."[413]

For Roberts, as for the majority of Egyptologists, Hathor is to be identified with the sun. But if Hathor represents the sun, how are we to understand her intimate relationships with Horus or Re, both of whom are typically identified with the sun by Egyptologists?

It can be shown, in fact, that Hathor has nothing whatsoever to do with the sun. As intimated in the previous chapter, the Eye-goddess Hathor is to be identified with the planet Venus.

The Eye of Horus

Of all the divine entities in the Egyptian pantheon, the Eye of Horus remains the most enigmatic and misunderstood. This is only to be expected, perhaps, given the fact that the original celestial identifications of Hathor and Horus continue to elude Egyptologists.

In order to make sense of the manifold symbolism attached to the Eye of Horus, it is essential at the outset of our investigation to come to grips with the Egyptian traditions telling of its incendiary rampage that purportedly brought

---

[410] M. Rikala, "Sacred Marriage in the New Kingdom of Ancient Egypt," in M. Nissinen & R. Uro eds., *Sacred Marriages* (Winona Lake, 2008), p. 122.
[411] J. Wilson, "Myths, Epics, and Legends," in J. Pritchard ed., *Ancient Near Eastern Texts Relating to the Old Testament* (Princeton, 1969), p. 11.
[412] J. Allen, *Middle Egyptian* (Cambridge, 2014), p. 36.
[413] A. Roberts, *Hathor Rising* (Devon, 1995), p. v.

the world to the very brink of extinction.[414] This tradition is most familiar, perhaps, from a text known as *The Destruction of Mankind*, inscribed on the tomb-walls of various kings from the 19th Dynasty and, as such, one of the oldest mythological narratives to survive from ancient Egypt.[415] There we read that the Eye, as the goddess Hathor, was dispatched by Re to punish mankind:

> *"Then mankind plotted something in the (very) presence of Re...Then they [Re's advisors] said in the presence of his majesty: 'May thy Eye be sent, that it may catch for thee them who scheme with evil things... It should go down as Hathor.' So then this goddess came and slew mankind in the desert."*[416]

The destructive campaign waged against mankind was commonly mythologized as a "bloodbath" (*ḥrt, šbbw*) wrought by the warrior-goddess Hathor or her alter ego Sakhmet. Indeed, in that same text we read: "Hathor will wade in the blood of mankind as Sakhmet."[417] The testimony of this text, coupled together with that from the earlier Pyramid and Coffin Texts, confirms that the Eye's rampage occurred in *zp tpj*—i.e., during the Time of Beginning when the ancient celestial gods still ruled on earth.[418]

Although *The Destruction of Mankind* dates to the New Kingdom,[419] the myth of the raging Eye-goddess is attested already in the pyramid of Unis (circa 2350 BCE) and was alluded to repeatedly throughout the three thousand years of Egyptian history. In Unis's pyramid we read that the flame from Horus's Eye produced an all-encompassing storm shaking the very foundations of heaven:

> *"I [Horus] will put flame in my eye, and it will encompass you and set storm among the doers of (evil) deeds, and its fiery outburst among these primeval ones. I will smite away the arms of Shu which support the sky."*[420]

---

[414] See J. G. Griffiths, "Remarks on the Mythology of the Eyes of Horus," *Chronique d'Egypte* 33 (1958), pp. 182-193; K. Sethe, *Zur Sage vom Sonnenauge* (Leipzig, 1912); G. Rudnitzky, *Die Aussage über 'Das Auge des Horus'* (Kopenhagen, 1956); R. Anthes, "Das Sonnenauge in den Pyramidentexten," *ZÄS* 86 (1961), pp. 1-21.

[415] W. Simpson, *The Literature of Ancient Egypt* (New Haven, 2003), p. 289.

[416] J. Wilson, "Deliverance of Mankind from Destruction," in J. Pritchard ed., *The Ancient Near East* (Princeton, 1958), p. 4.

[417] A. Spalinger, "The Destruction of Mankind...," *Studien zur Altägyptischen Kultur* 28 (2000), p. 291.

[418] A. Spalinger, *op. cit.*, p. 261. See also E. Hornung, "Exploring the Beyond," in E. Hornung & B. Bryan eds., *The Quest for Immortality* (New York, 2002), p. 33.

[419] S. Quirke, *Ancient Egyptian Religion* (London, 1992), p. 164 dates it to the time of King Tut. See also A. Spalinger, *op. cit.*, p. 259, who observes: "We are forced to date this literary narrative to Dynasty XVIII *at the earliest.*"

[420] *PT* 298-299.

So, too, the Coffin Texts are replete with scattered allusions to the apocalyptic disaster associated with the Eye of Horus. Spell 316 emphasizes the terrifying power of the raging goddess:

"I am the fiery Eye of Horus, which went forth terrible, Lady of slaughter, greatly awesome, who came into being in the flame of the sunshine, to whom Rē' granted appearings in glory…What Rē' said about her: Mighty is the fear of you, great is the awe of you, mighty is your striking-power…all men have been in the sleep of death because of you and through your power."[421]

The Eye of Horus is here likened to a heaven-spanning "flame" (*ns*). In Spell 946, as elsewhere, the Eye is associated with fire falling from the sky: "I am a fire in sky and earth, and all my foes are under my flame."[422] Other hymns report that the Eye's rampage was pyrotechnic in nature: "The fire will go up, the flame will go up…the fiery one will be against them as the Eye of Rē'."[423] The celestial context of the warring Eye-goddess is repeatedly emphasized: "Its flame is to the sky."[424]

At one time or another, every major Egyptian goddess is identified with the Eye of Horus: Hathor, Isis, Wadjet, Mut, Wepset, etc. The traditions attached to Sakhmet aptly illustrate the fundamental nature of the warring Eye-goddess. In the Bremner-Rhind papyrus it is Sakhmet who protects the king and wards off his enemies as the raging Eye:

"Thou art (condemned) to this fire of the Eye of Rē'; it sends forth (?) its fiery blast against thee in this its name of Wadjet; it consumes thee in this its name of 'Devouring Flame'; it has power over thee in this its name of Sakhmet; it is fiery against thee in this its name of 'Glorious Serpent'."[425]

The image of Sakhmet as a raging goddess is abundantly attested among the religious inscriptions discovered at Philae. Here, as elsewhere, Sakhmet is identified as the "Eye of Horus":

"Sakhmet, the strong one (wsrt), is in Bigeh in her form as the Eye of Horus, the living [eye…] while [spreading fire (?)] with the flame when

---

[421] *CT* IV: 98-100. Unless indicated otherwise, all quotations from the Coffin Texts are from R. Faulkner, *The Ancient Egyptian Coffin Texts* (Oxford, 2004).

[422] VII:162.

[423] *CT* V: 264.

[424] *CT* III:343.

[425] R. Faulkner, "The Bremner-Rhind Papyrus— IV," *Journal of Egyptian Archaeology* 24 (1938), p. 45.

she goes round, while scorching the rebels with the heat of her mouth. She is the primeval snake (krḥt)."[426]

In the temple texts from Edfu, Sakhmet is once again compared to a flame-throwing serpent and celebrated for her protective powers. Witness the following passage:

"*O Sekhmet, Eye of Rēʿ, great of flame, Lady of protection who envelops her creator...O Sekhmet who fills the ways with blood, Who slaughters to the limits of all she sees, Come towards the living image, the living Hawk, Protect him, and preserve him from all evil.*"[427]

According to scattered statements preserved in the Pyramid and Coffin Texts, the raging Eye-goddess was eventually pacified or otherwise dissuaded from her terrifying campaign of destruction. The following account from the Coffin Texts alludes to this motif: "The storm of Her who is mighty of dread, Mistress of the land, is quelled(?)."[428] The "pacification" of the raging Eye, in turn, is typically attributed to the heroic interventions of either Shu or Onuris (the two gods were commonly identified). In Spell 325 from the Coffin Texts, for example, Shu is described as follows: "He subdued the Eye when it was angry and fiery, that he might lead the Great Ones and have power over the gods...."[429] The same idea is alluded to in Spell 75, wherein Shu is made to announce:

"*I have extinguished the fire, I have calmed the soul of her who burns, I have quietened her who is in the midst of her rage...(even she) the fiery one who severed the tresses of the gods.*"[430]

Numerous texts from the Ptolemaic period likewise allude to the pacification of the raging Eye by Onuris or Shu, often aided and abetted by Thoth. Thanks in large part to the painstaking detective work and incisive scholarship of Herman Junker, these widely scattered fragmentary texts were pieced together and have since come to be known by the generic name of the *Wandering Goddess* theme.[431] Although they were inscribed nearly two thousand years later than the Pyramid and Coffin Texts, such texts frequently preserve archaic motifs and are thus still worthy of serious study—this despite the fact that they also betray rather clumsy attempts to historicize and localize the myth of the raging Eye goddess, originally celestial in nature.

---

[426] Quoted from J. F. Borghouts, "The Evil Eye of Apophis," *Journal of Egyptian Archaeology* 59 (1973), p. 136.
[427] Quoted from A. Roberts, *Hathor Rising* (Devon, 1995), p. 13.
[428] *CT* VII:36.
[429] *CT* IV:154.
[430] *CT* I:378.
[431] H. Junker, *Der Auszug der Hathor-Tefnut aus Nubien* (Berlin, 1911).

The basic plot of the *Wandering Goddess* myth finds the Eye goddess (variously identified as Hathor, Tefnut, Wadjet, Mut, etc.) going into exile and abandoning Egypt for some distant land—typically Nubia or Libya—whereupon she goes on a destructive rampage in the form of a raging eye or lion. It is only through the magical interventions of Shu, Onuris, or Thoth that the warrior-goddess is eventually pacified and induced to return to Egypt.

The myth of the *Wandering Goddess* as raging Eye formed a recurring theme in various New Year's rituals celebrated throughout Egypt during the Ptolemaic period, the latter being characterized by bouts of drinking, boisterous music, the brandishing of torches, and ecstatic dancing.[432] According to John Darnell, the clamorous music and dancing was thought to soothe the goddess and ward off noxious elements: "Their noisy revelry [of the celebrants] appeasing the ever more calm goddess, and the cacophony driving away baleful influences."[433]

A measure of the ritual's ideological importance can be seen from the fact that Egyptian monuments from this period depict the pharaoh emulating (or reenacting) the role of Shu/Onuris by performing a series of gyrating dances and other magical acts, such as the vigorous rattling of sistrums.[434] Such mimetic performances were evidently performed in the belief that they would help calm (*shtp*) the Eye's raging (*nšn*).

Although the pacification of the Eye of Horus is never spelled out as explicitly as we might wish, it seems clear that, upon being calmed, the Eye was returned or otherwise restored to the god Horus.[435] Indeed, a wealth of evidence suggests that the Eye-goddess eventually came to adorn the Horus-star as his royal headband or crown. An exemplary text in this regard is Spell 220/221 from the Pyramid Texts, quoted earlier, wherein the Eye-goddess is addressed by a series of epithets identifying her as the Red Crown (*Nt*):

"He has come to you, O Nt-crown; he has come to you, O Fiery Serpent; he has come to you, O Great One; he has come to you, O Great of Magic, being pure for you and fearing you...How kindly is your face, for you are content, renewed, and rejuvenated, even as the father of the gods fashioned you. He has come to you, O Great of Magic, for he is Horus encircled with the protection of his Eye, O Great of Magic...Ho, Crown great of magic! Ho Fiery Serpent! Grant that the dread of me be

---

[432] B. Richter, "On the Heels of the Wandering Goddess," in M. Dolinska &. H. Beinlich eds., *Ägyptologische Tempeltagung* (Wiesbaden, 2010), p. 155 writes: "one of the most important festivals celebrated during the Ptolemaic Era."

[433] J. Darnell, "Hathor Returns to Medamud," *Journal of Egyptian Archaeology* 22 (1995), pp. 92-93.

[434] H. Junker, *op. cit.*, pp. 9, 72, 85.

[435] J. G. Griffiths, *op. cit.*, p. 29.

like the dread of you; Grant that the fear of me be like the fear of you...
If Ikhet the Great has borne you, Ikhet the Serpent has adorned you;
If Ikhet the Serpent has borne you, Ikhet the Great has adorned you,
Because you are Horus encircled with the protection of his Eye."[436]

As stated here in no uncertain terms, it is the "encircling" (šn) of Horus by the fire-spewing uraeus-goddess—addressed as Ikhet the Serpent, "Great of Magic," and other archaic epithets—which provides the star-god with his "Eye" and thereby equips him with a magical rampart of protection (s3). Essential to understanding the symbolism involved is the report that the raging Eye is "calmed" (sḥtp) upon returning to and encircling the Horus-star as a crown—hence the Eye's description as "content" (ḥtp) in PT 195c.

The same basic idea is evident in Haremhab's coronation hymn, wherein we read that the uraeus-serpent—explicitly addressed as "Great of Magic"—was deemed responsible for the "crowning" of the Egyptian pharaoh: "[her arms] in welcoming attitude, and she embraced his beauty and established herself on his forehead, and the Divine Ennead…were in exultation at his glorious rising."[437]

So, too, a much later hymn from the Ptolemaic hemispeos at Elkab celebrates the return of the Wandering Eye-Goddess with the following words:

"Welcome! Says Re, 'Welcome! Come back upon the head of him whom
you have protected, the head from which you went forth!'"[438]

The return of the raging Eye to its rightful place atop the brow of the Horus-star, in addition to providing the royal crown, imbues the latter with awe-inspiring glory and power. Indeed, it is no exaggeration to say that the Eye-Goddess "invests" the Horus-star as the King of the Gods through this act of encirclement or "conjunction." Susan Johnson summarized the primeval mythological events in question as follows:

"The god of creation appeased the eye, which had become a cobra, by
placing it on his forehead as the uraeus, i'rt [Iaret], 'the Risen One',
who guards the crown. The pacification of the cobra thus marked the
establishment of monarchy, and the uraeus became the protective
symbol of legitimate kingship and unity."[439]

Johnson's conclusion is right on the money and bears emphasizing: It is the joining together or *reunion of the Eye/uraeus with the Horus-star* which marks

---

[436] PT 194-198.
[437] A. Gardiner, "The Coronation of King Ḥaremḥab," *Journal of Egyptian Archaeology* 39 (1953), p. 29.
[438] Quoted from J. Darnell, *op. cit.*, p. 50. J. Quack, "A Goddess Rising 10,000 Cubits into the Air," in J.M. Steele ed., *Under One Sky* (Münster, 2002), p. 288 notes: "It is frequently stated that the goddess comes as a head-ornament, or a snake on the brow of her father."
[439] S. Johnson, *The Cobra Goddess of Ancient Egypt* (London, 1990), p. 6.

the establishment of kingship.[440] The pacification of the raging Eye, moreover, signals the restoration of world order after the terrifying events attending its destructive rampage.[441]

Even from this brief survey it is evident that the Eye of Horus plays a central role in ancient Egyptian conceptions regarding kingship and divinity. How, then, are we to explain the peculiar mythological traditions attached to the raging Eye-goddess? James Allen, together with the vast majority of Egyptologists, would identify the Eye of Horus with the Sun:

"The uraeus-goddess is essentially the destructive power of the sun. She is a goddess because the sun is viewed in this case as the eye of the sun-god (Horus or Re), which is feminine (jrt). She is normally represented as a cobra (wadjet, also feminine) because of the notion of a power that can strike and kill: the hieroglyphic representation of this is N6 (sun-disk with cobra) [f]. The same power is viewed as inherent in the king's headgear, which is why it also has a uraeus. The primary thing to keep in mind, however, is the notion of the eye. In myth she represents both the sun as an eye and its destructive power. Her common epithet 'Great of Magic' (weret-hekau) derives from the idea of the eye as conveyor of intent—same notion as the 'evil eye' (which also existed in ancient Egypt). Through syncretism, she is associated with other major goddesses, such as Isis and Hathor."[442]

Yet if the Eye of Horus is to be identified with the Sun, how are we to understand its departure from Horus (or Ra in the *The Destruction of Mankind*)? For if Horus himself is to be identified with the Sun, as Allen assures us, one is left with the seemingly paradoxical situation wherein the Eye/Sun departs and, after threatening the world with destruction, returns to encircle or "crown" *itself*![443]

It must also be asked why Egyptian skywatchers would conceptualize one and the same celestial body as simultaneously both male (Horus) and female

---

[440] K. Goebs, *Crowns in Egyptian Funerary Literature* (Oxford, 2008), p. 143 has recently offered a similar assessment: "The return of the Eye(s) corresponds with the investiture of the new Horus-king."

[441] R. T. Clark, *Myth and Symbol in Ancient Egypt* (London, 1959), p. 94 wrote: "The return of the Eye marks the assumption of kingship by the High God and the end of the age of inchoate chaos." B. Richter, *op. cit.*, p. 156, wrote: "With her pacification, the order of the cosmos is also restored."

[442] Personal correspondence, 8-29-2010.

[443] J. Quack, "A Goddess Rising 10,000 Cubits into the Air," in J.M. Steele ed., *Under One Sky* (Münster, 2002), p. 287 offered a similar opinion: "As the goddess is always designated as the daughter of the sun-god to whom she returns, any interpretation identifying the goddess herself with the sun runs into serious trouble: one sun would have to play two roles at once."

(the Eye as Hathor or Sakhmet). Insofar as the Horus-star represented the celestial prototype of the masculine warrior-king, while the Eye represented the archetypal mother goddess, this interpretation of the Egyptian myth would appear to be at odds with the facts and more than a little strained.

The explicit catastrophic imagery attending the raging of the fire-spewing Eye is equally difficult to explain by the solar hypothesis. Under what circumstances does the Sun threaten to destroy mankind or the world? In what sense is the Sun displaced from heaven while presenting a terrifying fire-spewing serpentine form? What is there about the Sun's familiar appearance that would lead it to be conceptualized as a serpentine-goddess "protecting" the Horus-star by encircling it as a luminous crown or headband?

Far from being a reference to one and the same solar orb, Horus's relationship to the Eye is best understood as reflecting a close interaction or conjunction between two entirely different celestial bodies.[444] On this matter the preponderance of evidence is unequivocal: The Eye of Horus is to be identified with the mother goddess and, as such, it represents an entirely different star from that associated with the masculine Horus (Mars). In ancient Egypt, as around the globe, only one star fits the bill as the celestial prototype for the mother goddess—namely, Venus.[445] In this conclusion we find ourselves in complete agreement with Rolf Krauss who, in a number of publications investigating Egyptian astral religion, has presented evidence that the Eye of Horus is to be identified with the planet Venus.[446]

Granted the possibility that the planet Venus is the subject of the Egyptian traditions regarding the Eye of Horus, how are we to understand the cataclysmic imagery attending its destructive rampage? Whence derives the Eye's pronounced capacity for raining fire and destruction on mankind? Why would the planet Venus be conceptualized as a raging serpent at one moment and as "pacified" or calmed on another? On these all-important questions Krauss had nothing substantive to offer, noting simply: "It remains unclear how the observer understood raging and peacefulness."[447]

---

[444] R. Krauss, "The Eye of Horus and the Planet Venus: Astronomical and Mythological References," in J. Steele & A. Imhausen eds., *Under One Sky* (Münster, 2002), p. 194 reached a similar conclusion: "If this myth reflects reality or nature, then the eye which leaves the sun god and afterwards returns to him cannot be identical with the sun disk itself."

[445] E. Cochrane, *The Many Faces of Venus* (Ames, 2001), pp. 7-13; 159-165.

[446] R. Krauss, *Astronomische Konzepte und Jenseitsvorstellungen in den Pyramidentexten* (Wiesbaden, 1997), pp. 193-208. It will be noted that Talbott and I offered this identification well over a decade before Krauss.

[447] *Ibid.*, p. 201.

The ancient Egyptian texts provide a wealth of testimony on each of these questions. This testimony, in turn, can then be compared with literary traditions from ancient Mesopotamia and elsewhere that describe the planet Venus in analogous terms—i.e., as a terrifying agent of apocalyptic disaster, one prone to assuming a fire-spewing, serpentine form.[448] It was in the latter guise, according to the archaic traditions preserved in the Sumerian text known as *The Exaltation of Inanna*, that the war-mongering planet-goddess rained fire from heaven:

"Like a dragon you have deposited venom on the land, When you roar at the earth like Thunder, no vegetation can stand up to you. A flood descending from its mountain, Oh foremost one, you are the Inanna of heaven and earth! Raining the fanned fire down upon the nation...When mankind comes before you In fear and trembling at your tempestuous radiance."[449]

Virtually identical imagery is to be found in early literary accounts of the Semitic goddess Ishtar who, like the Sumerian Inanna, was explicitly identified with the planet Venus. The following passage is representative in this regard:

"I rain battle down like flames in the fighting, I make heaven and earth shake (?) with my cries...I constantly traverse heaven, then (?) I trample the earth. I destroy what remains of the inhabited world."[450]

Here, as in the Sumerian hymns describing Inanna, *it is the planet Venus* that is raining fire and destruction from the sky. The celestial context of the imagery in question could hardly be more explicit or unequivocal.

In ancient Mesopotamia, as in Egypt, the raging planet-goddess brings an apocalyptic storm in her wake. The testimony of a Sumerian text known as BM 23820 is instructive in this regard:

"Inanna, who pours down rain over all the lands, over all the people, loud-thundering storm. Hierodule, who makes heavens tremble, who makes the earth quake, Who can soothe your heart? You who pour down firebrands over the earthly orb, who flash like lightning over the highland...Whose cry reaches heaven and earth, whose roar is all-destructive...Your angry heart is a terrifying flood-wave."[451]

In this hymn, as in various other early Sumerian texts, Inanna/Venus rains down fire and flood from the skies much as was reported of the raging Eye

---

[448] See here the discussion in E. Cochrane, *On Fossil Gods and Forgotten Worlds* (Ames, 2010), pp. 77-123.
[449] W. Hallo & J. van Dijk, *The Exaltation of Inanna* (New Haven, 1968), pp. 15-17.
[450] B. Foster, *Before the Muses* (Bethesda, 1993), p. 74.
[451] S. Kramer, *From the Poetry of Sumer* (Berkeley, 1979), p. 89.

in Egyptian texts from the same general period (circa 2000 BCE). And as we found to be the case with the Eye of Horus, the Sumerian gods are described as desperate to pacify the planet-goddess's terrible rage. Indeed, the gods' attempt to dissuade the raging Inanna from her path of destruction is the subject of a rare and little-known text from the Old Babylonian period known as BM 29616. A few excerpts from this important hymn follow:

"Queen of Ibgal [Inanna], what has your heart wrought! How heaven and earth are troubled! What has your raging heart wrought! What has your flood-like raging heart wrought!"[452]

Even from this cursory survey the remarkable parallels between the mythological traditions describing Inanna/Venus and those attached to the Eye of Horus are readily apparent. Yet in the dozens of monographs devoted to the Egyptian Eye goddess—not to mention the hundreds of articles that have addressed the peculiar symbolism surrounding the raging Eye of Horus—I am not aware of a single scholar who has even commented on the manifold parallels between the Eye and Inanna/Venus, much less offered a systematic cross-cultural analysis of the traditions. As a result, modern Egyptology remains mired in the same long-outdated solar interpretations that dominated scholarship at the turn of the 20th century, wholly oblivious to the fact that ancient Egyptian descriptions of the raging Eye of Horus share a virtual one-to-one correspondence with Sumerian descriptions of the raging Inanna/Venus.[453]

To return to the Egyptian testimony, it will be remembered that the raging Eye produced a world-engulfing storm. In the following account from Unis's pyramid, quoted earlier, the flame from Horus's Eye is likened to a raging storm:

"I [Horus] will put flame in my eye, and it will encompass you and set storm among the doers of (evil) deeds, and its fiery outburst among these primeval ones. I will smite away the arms of Shu which support the sky."[454]

In addition to shaking heaven from its very moorings, the raging Eye unleashes a "storm" (*nšn*) on the evildoers. As for how we are to understand the "storm" in question, it is evident that it is a meteorological disturbance of cosmic proportions—in the Coffin Texts the word *nšn* is repeatedly used to denote an apocalyptic storm.[455] Yet in Spell 335 from the Coffin Texts the same

---

[452] S. Kramer, "BM 29616: The Fashioning of the *Gala*," *Acta Sumerologica* 3:3 (1981), p. 5.

[453] E. Cochrane, *On Fossil Gods and Forgotten Worlds* (Ames, 2010), pp. 77-123.

[454] *PT* 298-299.

[455] *CT* 4:40, 4:396, VI:306, and 7:376. According to R. Caminos, *The Chronicle of Prince*

word is employed to describe the "wrath" of the raging Eye-goddess: "I raised the hair from the Sacred Eye at its time of wrath."[456] The seemingly incongruous reference to "hair" (*šn*) in conjunction with a stellar Eye is addressed—if not fully clarified—by a gloss appended to the spell in question:

"What is the Sacred Eye at its time of wrath? Who raised the hair from it? It is the right Eye of Re when it was wroth with him after he had sent it on an errand. It was Thoth who raised the hair from it."[457]

Here the disaster-bringing "wrath" (*nšn*) of the raging Eye is explicitly linked to its "hair" (*šn*). Indeed, if we are to judge from this gloss, the raging Eye was only calmed or pacified with the "raising" of its hair. Taking these two traditions together, it seems evident that the "wrath" or "storm" associated with the Eye of Horus was somehow connected to its extraterrestrial "hair" (see figure 1).

Figure 1

The same conclusion is supported by the apparent etymological relationship between the Egyptian words *šnj*, "hair," and *šnjt*, "storm," the latter term being employed to denote the primeval storm that accompanied the deceased king's ascent to heaven (indeed this very term is elsewhere substituted for *nšn*).[458] The following passage is one of several that could be cited in this context: "N will

---

Osorkon (Rome, 1958), p. 90 *nšn* denotes "the rage of heaven" or a "convulsion of the sky."
[456] *CT* IV: 238.
[457] *CT* IV: 239-240.
[458] See especially *PT* 1150, 2366 and *CT* V:150, VI:330. See also the discussion in R. Faulkner, *The Ancient Egyptian Coffin Texts* (Oxford, 1973), p. 86.

dispel the storm."⁴⁵⁹ (It will be noted that the word translated as "dispel" here is <u>t</u>s, an apparent cognate of the word <u>t</u>s translated as "raised" in the passage above describing Thoth's raising of the hair from the Sacred Eye.)

Here, then, is the probable answer to the question which so baffled Krauss regarding the Egyptian traditions describing the Eye of Horus as the planet Venus—namely, how to explain its peculiar capacity for "raging": According to the express testimony of the Pyramid and Coffin Texts, the Eye/Venus was conceptualized as raging because it displayed disheveled "hair" while circling about the sky—this while raising a "storm" and spewing forth immense volumes of fiery material. The pacification of the Eye, on the other hand, has reference to a subsequent phase of planetary history during which Venus's "storm" was dissipated and its "hair" tied-up or otherwise brought under control ("uplifted" in the previous passage quoted from Spell 335).⁴⁶⁰

If this much is fairly obvious and straightforward, it remains to be shown how or why the planet Venus could ever be conceptualized as suddenly displaying disheveled or "storm"-laden hair. The answer to this question, it must be said, will not sit well with the Central Dogma of modern astronomy, which holds that the visible planets have not changed their appearance or orbits in any fundamental manner for many millions of years. Yet it is necessary to follow the evidence wherever it happens to lead. As Sherlock Holmes was wont to say, "When you have eliminated the impossible, whatever remains, however improbable, must be the truth."⁴⁶¹ As improbable as it must appear at first sight, the testimony of ancient skywatchers with regards to Venus's recent history tells a very consistent and compelling story—namely, that of a planet run amok.

In order to clarify the nature of Venus's "hair" from the standpoint of natural science, it is instructive to review the Egyptian traditions recounting how the Eye-goddess came to encircle the Horus-star with a royal head-band. Spell 510 from the Pyramid Texts is especially informative in this regard. There the Eye is addressed as Ikhet the Great: "I am this head-band of red colour which went forth from Ikhet the Great; I am this Eye of Horus which is stronger than men and mightier than the gods."⁴⁶² Here the all-powerful Eye of Horus is specifically identified with a head-band emanating from the Ikhet-Serpent. The head-band associated with the Eye is also alluded to in another equally

---

[459] *CT* VII:110.
[460] See here the valuable discussion in R. Nyord, *Breathing Flesh: Conceptions of the Body in the Ancient Egyptian Coffin Texts* (Copenhagen, 2009), pp. 228-230.
[461] *The Sign of the Four*.
[462] *PT* 1147.

enigmatic passage from Spell 519: "So that I may ferry across in it together with that head-band of green and of red cloth which has been woven from the Eye of Horus in order to bandage therewith that finger of Osiris that has become diseased."[463]

The Egyptian word translated as "head-band" in both of these Spells is *sšd*, determined with the following glyph: ⌒. If Horus is to be identified with a celestial body, as all Egyptologists agree, how are we to understand this encircling head-band?[464] This question is directly related to another: Why would a "head-band" be likened to—or identified with—a terrifying serpentine form which rains fire and "is stronger than men and mightier than the gods"?[465]

A decisive clue is provided by the fact that, in addition to denoting "head-band" the word *sšd* also signifies a flame-scattering star or comet-like object.[466] Indeed, there are a number of Egyptian texts which describe an *sšd*-star streaking across the sky and scattering fire. In the so-called Poetical Stela found at Karnak, for example, the warring Thutmose III is likened to an *sšd*-star "strewing its fire in flame and yielding its downpour."[467] As Wainwright pointed out in his discussion of the royal inscription in question, the terrifying specter of the *sšd*-star is a decided point of emphasis in the Egyptian texts: "Its dangerous nature is certified by the desire of the Pharaoh to seem to his enemies in battle to be like the *sšd*."[468]

Hitherto overlooked, however, is the fact that the *sšd*-star's propensity for strewing flames of fire and inspiring dread is mirrored in Egyptian reports

---

[463] *PT* 1202-1203.

[464] C. Manassa, *The Late Egyptian Underworld* (Wiesbaden, 2007), p. 26 suggests that the reference is to the Milky Way: "The *sšd*-fillet in this Pyramid Text passage [Spell 519] may also allude to the Milky Way, providing further evidence for the Egyptian association of the eye of Horus with astral bodies other than the sun and moon."

[465] It will be noted that the *sšd*-headband is also identified with the Eye of Horus in PT 96c.

[466] R. Hannig, Ägyptisches Wörterbuch I (Mainz, 2003), p. 1244, entry 30768. J. Zandee, *Der Amunhymnus des Papyrus Leiden I 344, Verso, Vol. 1* (Leiden, 1992), p. 356 translates *sšd* as "Komet."

[467] Line 15 of the king's Poetical Stela as translated in R. Faulkner, "'The Pregnancy of Isis': A Rejoinder," *Journal of Egyptian Archaeology* 59 (1973), p. 219, citing Urk. IV, 615, 13-15. K. Goebs, *Crowns in Egyptian Funerary Literature* (Oxford, 2008), p. 373 translates the passage as "a shooting star that scatters its flame of fire."

[468] G. Wainwright, "Letopolis," *Journal of Egyptian Archaeology* 18 (1932), p. 162. R. Faulkner, *op. cit.*, p. 219 translated *sšd* as "lightning-flash," arguing "such a description is not applicable to a star, but is most appropriate to a thunderstorm." The fact that *sšd* is determined with a star in Thutmose's stela undermines Faulkner's strained interpretation. Note also the fact that *sḏt*, denoting the fire strewn by the *sšd*-star, is almost certainly cognate with *sḏ/sd*, denoting "tail," a likely reference to a comet-like object (the latter word is determined with the "hair" glyph).

of the raging Eye of Horus. Recall again the passage from Unis's pyramid, quoted earlier, wherein the flame from Horus's Eye is likened to a raging storm: "I [Horus] will put flame in my eye, and it will encompass you and set storm among the doers of (evil) deeds, and its fiery outburst among these primeval ones."[469]

The Eye of Horus is here said to unleash a fiery "outburst" or ($hfhft$) on the primeval ones.[470] This tradition recalls the fact that a "downpour" of fiery efflux (jdt) was associated with the $sšd$-star on Thutmose's Poetical Stela.[471] How, then, are we to explain the Eye's peculiar association with a flood of fire?

That there was a celestial basis for such imagery is strongly suggested by the fact that an extraterrestrial "flooding" and raining of fire is precisely the disaster dispensed by Inanna/Venus, as evidenced not only by the Sumerian story quoted above (BM 23280) but also by the following hymn as well:

"Like a dragon you have deposited venom on the land...A flood descending from its mountain, Oh foremost one, you are the Inanna of heaven and earth! Raining the fanned fire down upon the nation."[472]

In light of the striking parallels between Egyptian traditions describing the Eye of Horus and Sumerian hymns celebrating Inanna/Venus, it is necessary to ask whether the two celestial bodies simply shared a fundamental affinity or were conceptually related in some fashion? On this point the evidence is unequivocal: The Eye of Horus, like Inanna, represents the planet Venus. It was the Queen of Heaven that was conceptualized as a raging warrior-goddess throughout the ancient Near East (Inanna, Ishtar, Astarte, Anat, etc.), thereby paralleling the behavior accorded the Eye/Hathor in Egyptian texts.[473] And it was the planet Venus that formerly presented a comet-like form and scattered fire across the sky, an event mythologized as a fire-scattering "comet" ($sšd$) or as a fire-spewing serpent (the Inanna-dragon or uraeus-goddess).[474] The raging "storm" associated with the fire-spewing Eye of Horus, in turn, finds

---

[469] *PT* 298-299.
[470] R. Faulkner, *op. cit.*, p. 190 translates the word as "flood." The very same word is used to describe a flood deposited by the Eye of Horus in the Coffin Texts (VII:235h).
[471] *Jdt* elsewhere denotes the "wrath" of the Eye of Horus. See *CT* IV:108.
[472] W. Hallo & J. van Dijk, *The Exaltation of Inanna* (New Haven, 1968), pp. 15-17. See also line 29 from "A hymn to Inana (Inana C)," *ETCSL* wherein Inanna is invoked as follows: "Her wrath (is)...a devastating flood which no one can withstand."
[473] It will be noted that the Egyptian *Astarte Papyrus* describes the planet-goddess Astarte as *qndt nšny*, "furious/raging storm." See here the discussion in N. Ayali-Darshan, "The Other Version of the Story of the Storm-god's Combat with the Sea...," *JANER* 15 (2015), p. 33.
[474] E. Cochrane, *The Many Faces of Venus* (Ames, 2001), pp. 113-151. See also D. Talbott, "The Great Comet Venus," *AEON* 3:5 (1994), pp. 5-51.

a precise structural analogue in the raging "storm" (UD.HUŠ) associated with the incendiary rampage of Inanna/Venus[475] and has direct reference to the terrifying meteorological phenomena associated with the disheveled "hair" of the Venus-comet, alleged to have spanned the sky and eclipsed the Sun.[476]

To summarize our findings in this chapter: A wealth of evidence confirms that the Eye of Horus is to be identified with the planet Venus. This finding, in turn, has profound and wide-ranging ramifications for the proper understanding of ancient Egyptian religion insofar as the Eye's history is catastrophic from start to finish, being explicitly associated with apocalyptic cosmic disaster *in illo tempore*. After an indefinite period of time during which it was conceptualized as waging war amidst darkness and disorder, the raging Eye was eventually pacified or "calmed," whereupon order was restored to the cosmos. According to the archaic account preserved in Spells 220-221 from the Pyramid Texts, the Eye eventually came to encircle the star-god Horus as a serpentine goddess (Ikhet the Serpent), thereby providing him with an aegis-like head-band or crown. The Eye-goddess's "head-banding" of the Horus-star, in our view, has direct reference to extraordinary astronomical events in which the planet Venus appeared to conjoin with Mars (Horus) and encircle it with a comet-like band (*sšd*)—hence the reason why this one word denotes at once Ikhet's head-band as well as a flame-scattering star ("meteor or comet"). Indeed, the mere fact that the word *sšd* is elsewhere determined by the following glyph—f— confirms its intimate relationship with the uraeus-serpent that came to form the encircling crown of Horus.[477]

---

[475] See lines 204-205 of "The death of Ur-Nammu (Ur-Namma A)," *ETCSL*.
[476] See here the discussion in E. Cochrane, *On Fossil Gods and Forgotten Worlds* (Ames, 2010), pp. 77-113.
[477] W. Barta, "Zur Bedeutung des Stirnbands-Diadems," *Göttinger Miszellen* 72 (1984), p. 8.

13. The Latin Goddess Venus

> *"Who will deny that, of all the immortals, Venus is the most baffling?"*[478]
> *"And the force that brings their vinctio 'binding' is Venus 'Love.'"*[479]
> *"Just tie them and repeat 'the Venus knot I tie.'"*[480]

The first temple to Venus was vowed in 295 BCE, in the aftermath of a bitter and costly campaign against the Samnites. Prior to that occasion the origins of the Latin goddess and her cult are obscure, to say the least.[481]

There would appear to be general agreement among scholars that the goddess Venus is simply the personification of a magical force originally meaning something like "charm."[482] It was Robert Schilling who, in his monumental *La Religion romaine de Vénus*, pointed the way to a new understanding of the Latin goddess. There he analyzed a family of words deemed to be cognate with Venus: *venia, venerari, venenum, venenatum,* and *venerium*. Each of these words, according to Schilling's interpretation, originally bore the meaning of "to charm," "to enchant," "to cast a spell," etc. Schilling maintained that this vocabulary takes us back to a religion saturated with magic, one which he claimed could properly be called Venusian:

> *"The neuter venus is part of a remarkable semantic series of the same kind as genus/Genius/generare, except that here the first term and not the second was divinized, passing from the neuter to the feminine: Venus/venia/venerari (sometimes venerare in Plautus). To the persuasive charm that the goddess embodies and that the venerans ('he who venerates') practices upon the gods, there corresponds the symmetric notion of venia in the sense of 'grace' or 'favor'—a notion that belongs*

---

[478] W. D. Anderson, "Venus and Aeneas," *Classical Journal* 50:5 (1955), p. 233.
[479] V 61 from M. Varro, *On the Latin Language* (Cambridge, 1967), p. 59.
[480] Virgil, *Eighth Eclogue* 78.
[481] J. B. Rives et al, "Venus," in H. Cancik & H. Schneider eds., *Der Neue Pauly* 12:2 (Stuttgart, 2000), col. 17.
[482] J. Scheid, "Venus," in S. Hornblower & A. Spawforth eds., *The Oxford Classical Dictionary* (Oxford, 1996), p. 1587: "The debate over the original nature of this goddess, who does not belong to Rome's oldest pantheon but is attested fairly early at Lavinium, has been partly resolved (Schilling 1954). It is now accepted that the neuter *venus*, 'charm', cannot be separated from the terms *venia, venerari, venenum* ('gracefulness, to exercise a persuasive charm', 'poison')."

to the technical vocabulary of the pontiffs (Servius, Ad Aeneidem 1.519). This metamorphosis of a neuter noun into a goddess...was very likely furthered by the encounter of this divinity with the Trojan legend. This legend must have facilitated the relation drawn between a Venus embodying charm in its religious meaning and an Aphrodite personifying seduction in the profane sense."[483]

Schilling's contribution to the debate over the Latin goddess's origins is undeniable and in recent years his theory seems to have gained near-universal acceptance within the scholarly community. Hendrik Wagenvoort, for example, has stated: "It is today an almost general opinion that the goddess' name evolved out of a neuter noun *venus*, *veneris*, indicating some sort of occult force."[484]

It remains the case, however, that while Schilling may have shed some much-needed light on the etymology behind the term *venus*, the origins of the Latin goddess remain as obscure as ever. Certainly it must be a rare phenomenon in the history of religion for so complex a goddess as the Latin Venus to evolve from an impersonal "force" employed in incantations and prayers. Equally elusive is the phenomenological basis behind the magical power to charm or enchant associated with Venus's name. Was it simply a mystical or spiritual act on the part of the worshipper, as envisaged by Schilling? Or did Venus's ability to "charm" or enchant have an empirical or material basis in the natural world? Wagenvoort expressed similar reservations about Schilling's thesis: "But if Venus was a 'certain mysterious force', as Schilling says—and I do not disagree with him—whatever can we suppose it to have been?"[485]

It is our opinion that there is a perfectly straightforward explanation for Venus's ability to charm or enchant—namely, her ability to bind or encircle with bonds or knots. In order to demonstrate the point, it is necessary to offer a brief survey of magic in ancient cultures.

On the Origins of Magic

It is well known that the act of binding forms an essential component of various forms of magic. On this matter Richard Onians observed:

"In many languages the general notion 'magic' is expressed by the term for 'binding'. It has been noted that it is the commonest process of bewitchment."[486]

---

[483] *La Religion romaine de Vénus* (Paris, 1954), p. 250.
[484] H. Wagenvoort, "The Origin of the Goddess Venus," in *Pietas* (Leiden, 1980), p. 168.
[485] *Ibid.*, p. 175.
[486] R. Onians, *The Origins of European Thought* (New York, 1973), p. 372.

The Assyrian word *abâru*, for example, signified "to bind" and to tie a magic knot as well as "spell" or charm.[487] The Greek *katadesmos* properly means "to bind," but it was frequently used with specific reference to the magic associated with knots and bands.[488] The most common term for magic amongst the ancient Norsemen—*seiðr*—is acknowledged to trace to a root meaning "band" or "fetter," with particular reference to the tying of magical knots.[489]

In ancient Egypt a common means of working magic was through the casting of bonds and the tying of knots. Joris Borghouts, in attempting a definition of magic, notes that "it is a moveable entity, not unlike a rope which may be knot, cast out, caught, and made to encircle something."[490] Notable here is the fact that the Egyptian word *šni* signifies "to enchant," as well as "to bind" or "encircle."[491]

The Latin language confirms that similar conceptions were current in ancient Roman times. Especially relevant is the word *vincula*, used by Horatius Flaccus in the sense of "to enchant," but clearly related to *vinculum*, "band."[492] It will be noted, moreover, that *vincula* and *vinculum* both appear to be derived from *vincio*, "to bind."[493]

Venus as Binder

It is tantamount to a truism in ancient religion that all great goddesses are "binders."[494] The Latin Venus is no exception in this regard: It was through the agency of binding, according to Marcus Varro, that she afflicted mortals and immortals alike with the pangs of love: "And the force that brings their *vinctio* 'binding' is Venus 'Love.'"[495] Indeed, Rome's greatest scholar speculated that it was Venus's ability to "bind" that provided the force behind the union of male and female.[496] Tibullus's claim that the goddess had bewitched him with a love-bond betrays a similar belief-

---

[487] F. Brown & S. Driver & C. Briggs, *A Hebrew and English Lexicon* (Oxford, 1951), pp. 287-288.
[488] H. Liddell & R. Scott, *A Greek-English Lexicon* (New York, 1872), p. 715.
[489] J. de Vries, *Altnordische etymologisches Wörterbuch* (Leiden, 1977), p. 488.
[490] J. F. Borghouts, "Magic," in *Lexikon der Ägyptologie* (Wiesbaden, 1972-1992), col. 1140.
[491] R. Ritner, *The Mechanics of Ancient Egyptian Magical Practice* (Chicago, 1993), p. 43. The related word *šnjt* signifies "encircling" as well as "conjuration."
[492] R. Onians, *op. cit.*, p. 268.
[493] M. de Vaan, *Etymological Dictionary of Latin and the other Italic Languages* (Leiden, 2008), p. 679.
[494] E. Neumann, *The Great Mother* (Princeton, 1955), p. 226-239.
[495] V 61 from M. Varro, *On the Latin Language* (Cambridge, 1967), p. 59.
[496] *Ibid.*

system: "Venus herself bound my arms with a magic knot."[497] Far from being a mere figure of speech, Venus's love-bond is best understood as an actual band as Richard Onians deduced.[498]

If the earliest religion of Venus is to be understood as saturated by the magical acts of charming, enchanting, and the casting of spells—as per Schilling—it stands to reason that such magic most likely involved the act of binding and/or the casting of bonds. Virgil, as we have seen, referenced Venus's capacity for knotting. Indeed, Europeans continued to speak of Venusian bands and knots many centuries after the eclipse of Roman religion.[499]

If the precise origins of the Latin goddess are lost in antiquity, it is generally agreed that her cult owes much to the influence of the Greek Aphrodite and analogous goddesses from the ancient Near East. Whether this influence is to be understood as direct in nature, or as a product of mere imitation, remains unclear. Georges Dumézil had this to say in *Archaic Roman Religion*:

"It may be imagined that feminine charm with its cunning approach, so powerful over its masculine objects, was designated by the same word [venus] as the captatio of the god by man. Hypothetical, to be sure, yet this explanation is the most likely that has been set forth so far. It is this *uenus which was personified, and in the feminine gender, which was particularly suited to signify all kinds of forces. Spontaneous evolution? An artifice to obtain for Rome and for the Latin vocabulary an equivalent of the Greek enchantress, Aphrodite or of her Etruscan shadow, Turan? Such an influence is more than likely...Thus, from the very first manifestation of Venus, one must think of the Foreign Goddesses, and ultimately of the Foreign Goddess, who gave rise to her. Actually, before the third century only one of her cults was known at Rome, under the name Calua, which concerns feminine charm in one of its aspects which for a long time was undisputed, the hair."[500]

Aphrodite herself is the Foreign Goddess *par excellence*, the ancient Greeks themselves tracing her cult to the ancient Near East.[501] This being the case, it stands to reason that additional insight into the origins of the Latin Venus might be achieved by examining the cult of the Sumerian goddess Inanna, with whom both Aphrodite and Venus share much in common.

---

[497] Tibullus 1, 8, 5 as translated in R. Onians, *op. cit.*, p. 373. The phrase in question is: "*ipsa Venus magico religatum bracchia nodo perdocuit.*"
[498] *Ibid.*, pp. 370-373.
[499] *The Compact Edition of the Oxford English Dictionary* (Oxford, 1971), pp. 3608-3609.
[500] G. Dumézil, *Archaic Roman Religion, Vol. 2* (Baltimore, 1966), p. 422.
[501] Pausanias, *Book* I:14:7.

Venus in Ancient Mesopotamia

The greatest goddess of the ancient world was the Sumerian Inanna (see chapter two). Like the Latin Venus, Inanna was especially beloved by kings, whom she accompanied in battle and led to victory. Sargon the Great (circa 2300 BCE) credited his success as a warrior to Ishtar/Irnina: "As long as Ištar gain [sic] victories for him…As long as Irnina will attain victories for him."[502] The epithet *Irnina* denotes "victory" and thus encodes the fact that Ishtar/Venus was the guarantor of victory on the battlefield.[503] The Latin Venus bears an analogous epithet: *Victrix*, signifying victory, a word derived from *vincio*, "to bind, tie."

The Exaltation of Inanna describes the goddess as a terrifying dragon raining fiery venom from the sky: "Like a dragon [ušumgal] you have deposited venom [uš$_{11}$] on the land."[504] Evident here is the conceptualization of Inanna as a venom-spewing dragon, a common image of the raging goddess around the world. It is to be noted that the Sumerian language preserves an explicit semantic link between the concepts "venom/poison" and "to cast a spell or charm"—the word uš$_{11}$ signifies "venom" as well as "poison, spittle, spell, charm."[505] The fact that the very same semantic link can be found in the Latin language—*venenum*=venom, poison, charm, to cast a spell—hints at a widespread (and archaic) association of ideas. If so, important questions arise as to how this particular datum is to be explained from the standpoint of natural science. Although the semantic link between "venom" and "poison" is readily understood by reference to a serpent's deadly venom, it is difficult to discern a logical link between a serpent's venom and the seemingly unrelated concepts of "spell" and "charm."

It is the celestial context of Inanna's mythology that points the way to a resolution of this archaic conundrum. As we have documented, the venom-spewing Inanna-dragon is indissolubly connected to the planet Venus. An early temple-hymn translated by Sjöberg and Bergmann confirms that Inanna—as the planet Venus—was indeed conceptualized as a dragon:

"Your queen (is) Inanna…the great dragon…Through her the firmament is made beautiful in the evening."[506]

Granted that the planet Venus, as Inanna, was identified as a serpent, how are we to explain the conceptual link between the dragon-goddess's venom

---

[502] H. Casanova, *Imagining God* (Eugene, 2020), p. 184.
[503] J. Westenholz, *Legends of the Kings of Akkade* (Winona Lake, 1997), p. 78.
[504] Line 9 as translated in W. Hallo & J. van Dijk, *The Exaltation of Inanna* (New Haven, 1968), p. 15.
[505] J. Halloran, *Sumerian Lexicon* (Los Angeles, 2006), p. 305.
[506] Å. Sjöberg & E. Bergmann, *The Collection of the Sumerian Temple Hymns* (Locust Valley, 1969), p. 36.

and a magical "spell" or "charm"? Given our earlier discussion with regards to the origins of magic, one is naturally led to suspect some sort of relationship between the serpent-like planet and the process of binding or encircling.

In ancient Egypt, where a venom-spewing serpent-goddess forms a prominent figure in the earliest pantheon, one can find a wealth of evidence which helps to clarify the testimony from ancient Mesopotamia. In light of our discussion thus far, it will be of paramount importance to discover whether the Egyptian uraeus-goddess is associated with the casting of magical spells and, if so, whether a connection to the planet Venus can be established.

The Uraeus-Goddess

The Egyptian mother goddess—alternately invoked as Isis, Sakhmet, and Wadjet, among other names—is consistently represented as a fire-spewing uraeus-serpent. As outlined in a previous chapter, the symbolism attached to the uraeus-serpent would appear to find its *raison d'être* in the singular events associated with the crowning of Horus. According to the archaic traditions preserved in the Pyramid Texts, the uraeus-serpent came to adorn the forehead of Horus as the crown of kingship during the decidedly catastrophic events attending Creation. A Pyramid Text devoted to the red crown (*Nt*-crown), wherein the uraeus-serpent is addressed as Ikhet, is instructive here:

"He has come to you, O Nt-crown; he has come to you, O Fiery Serpent; he has come to you, O Great One; he has come to you, O Great of Magic, being pure for you and fearing you...He has come to you, O Great of Magic, for he is Horus encircled with the protection of his Eye, O Great of Magic...Ho, Crown great of magic! Ho Fiery Serpent! Grant that the dread of me be like the dread of you; Grant that the fear of me be like the fear of you...If Ikhet the Great has borne you, Ikhet the Serpent has adorned you; If Ikhet the Serpent has borne you, Ikhet the Great has adorned you, Because you are Horus encircled with the protection of his Eye."[507]

As explicitly stated here in no uncertain terms, the great mother goddess, alternately identified as "Fiery Serpent" and "Eye," comes to encircle Horus and, as a result, provides him with the crown of kingship. A detailed analysis of the language in question will confirm that *it is the uraeus-goddess's act of binding or encircling Horus that constitutes the prototypical act of magic*. Thus, in the Spell in question the uraeus-serpent is said to "encircle" (*šnj*) Horus. Yet the same word means "to enchant."[508]

---
[507] *PT* 194-197.
[508] R. Hannig, *Ägyptisches Wörterbuch I* (Mainz, 2003), p. 1309.

Hence it can be surmised that, in encircling or "binding" Horus, the uraeus-serpent thereby enchanted him.

The fact that the uraeus-goddess was regarded as a terrifying and all-powerful celestial agent is essential to elucidating the multifaceted symbolism in question: By her act of encircling Horus as a serpent-like crown, the goddess was conceptualized as having provided him with an invincible means of magical protection and power. Here, then, is our answer to the conundrum posed earlier: i.e., how to explain the semantic link between the concepts "venom" and "poison" and magical "spell" or "charm." The original historical connection between the seemingly disparate concepts in question would appear to be explained as follows: It was the venom-spewing uraeus-goddess that came to encircle the god Horus as his crown, thereby "binding" and enchanting him while providing him with magical protection. Hence it is that the serpentine crown-goddess was known by the epithet "Great of Magic." The same epithet has been translated as "Great-in-charms"[509] and was commonly applied to Isis. Isis herself, moreover, was the Egyptian counterpart of the Latin charmer Venus. The fact that Isis was specifically identified with the planet Venus brings the argument full-circle.[510]

The epithet "Great of Magic" was also applied to the Eye of Horus, the latter entity being identifiable with the planet Venus.[511] Thus a pyramid text reads as follows: "Horus has put his eye on your brow in its name of Great-of-Magic."[512] As "Great of Magic/Charms," the Venusian Eye-goddess forms a close structural analogue to the Latin goddess Venus, mistress of charms.

Inanna/Venus and the Crown of Kingship

Given the striking parallels between the Egyptian and Mesopotamian accounts of the raging serpentine-goddess, it is relevant to ask whether Inanna/Venus, like the Egyptian Ikhet/Isis, was associated with the crown of kingship. In fact, it is Inanna who provides the crown in an early hymn: "To give the crown, the throne and the royal sceptre is yours, Inana."[513]

Yet there is much more that can be said with respect to Inanna's intimate relationship to the crown of sovereignty.[514] Thus, in a recent analysis of the

---

[509] S. Mercer, *The Pyramid Texts in Translation and Commentary*, Vol. II (New York, 1952), p. 93.
[510] Ptolemy, *Tetrabiblos* 2.3.64.
[511] R. Krauss, "The Eye of Horus and the Planet Venus: Astronomical and Mythological References," in J. Steele & A. Imhausen eds., *Under One Sky* (Münster, 2002), pp. 193-208.
[512] *PT* 1795.
[513] J. Black et al, *The Literature of Ancient Sumer* (Oxford, 2004), p. 96.
[514] See E. Cochrane, *Starf*cker* (Ames, 2006), pp. 155-158.

earliest pictograph used to write Inanna's name—the so-called MUŠ₃-sign (see figure 1)—Piotr Steinkeller suggested that it originally represented a headband or crown. Steinkeller concluded his article as follows:

"As for its specific meaning, we undoubtedly find here a type of band. Since suh is compared to 'crown' (men) and 'tiara' (aga), and since it may have been decorated with lapis lazuli, it certainly was an object of considerable importance and value, which was worn over the head. A translation 'diadem' would thus not be inappropriate…It would seem, therefore, that the archaic symbol of Inanna depicts a scarf or headband."[515]

Figure 1

In her magisterial monograph on Inanna's cult, Françoise Bruschweiler devotes several pages to an analysis of the Sumerian word MUŠ₃. There she notes that it served as a mark of sovereignty and was typically described as tied on or knotted (Sumerian kešda).[516] In the Sumerian hymn "Enki and the World Order," for example, the MUŠ₃ is mentioned together with the crown of sovereignty: "He raised a holy crown over the upland plain. He fastened a lapis-lazuli beard to the high plain, and made it wear a lapis-lazuli headdress [MUŠ₃]."[517] In the passage in question, the lapis-colored MUŠ₃ is set in apposition with the holy crown of sovereignty, thereby implying a direct connection between the tying on of the MUŠ₃ and the crown of kingship.

It is significant to note that, already at the dawn of history, Inanna/Venus was renowned for tying kingship on the Sumerian king. This idea is attested in a text from the Early Dynastic period (circa 2400 BCE), wherein the planet-goddess ties kingship on Lugalkiginnedudu: "When Inanna had tied the

---

[515] P. Steinkeller, "Inanna's Archaic Symbol," in J. Braun et al eds., *Written on Clay and Stone* (Warsaw, 1998), p. 95.

[516] F. Bruschweiler, *Inanna la déesse triomphante et vaincue dans la cosmologie sumérienne* (Leuven, 1988), pp. 124-125.

[517] Lines 349-350 from "Enki and the world order," *ETCSL*.

lordship with the kingship for Lugalkiginnedudu, she let him exert lordship in Uruk, she let him exert kingship in Ur."[518]

Egyptian tradition preserves a similar story about the goddess Isis. According to the Pyramid Texts, Isis tied a head-band or fillet on Horus at the time of Beginning: "Isis the Great, who tied on the fillet in Chemmis when she brought her loin-cloth and burnt incense before her son Horus the young child."[519] In the passage in question the word for "fillet" is *mdḥ*. The fact that this very word is employed to describe the king's headband or crown in the Coffin Texts suggests that it originally signified the crown of sovereignty. Witness the following passage:

"Come, that you may see me adorned with a fillet [mdḥ] and wearing the royal head-cloth. Joy is given to me by means of it."[520]

The most common ideogram or determinative for *mdḥ* is depicted in figure two.[521] The resemblance between this ideogram and the Sumerian MUŠ₃-sign used to denote Inanna/Venus is evident at once (figure 2). The fact that both ideograms were employed to describe the king's head-band or crown confirms their fundamental affinity and suggests that their spiraling volute-like form traces to a common celestial prototype.

Figure 2

In lieu of the possible historical connection between the cults of Sumerian Inanna, Greek Aphrodite, and Latin Venus, it is relevant to ask whether the latter goddess was associated with the royal crown? On this matter the evidence is unequivocal, as Varro attests:

"She [Venus] is connected with the corona, 'garland'...because the garland is a binder of the head and is itself, from vinctura, 'binding,' said vieri 'to be plaited,' that is, vinciri, to be bound."[522]

---

[518] P. Beaulieu, *The Pantheon of Uruk during the Neo-Babylonian Period* (Leiden, 2003), p. 105.
[519] *PT* 1214. Thus James Allen, *The Ancient Egyptian Pyramid Texts* (Atlanta, 2005), p. 161 translates the passage as "who tied the headband on her son Horus as a young boy."
[520] *CT* V:158.
[521] M. Betrò, *Hieroglyphics* (New York, 1996), p. 186.
[522] V 62, as quoted in Varro, *On the Latin Language* (Cambridge, 1967), pp. 59-61.

It will be noted that Varro understood the royal headband/garland to be something that was bound on.

The Shen Bond

The Egyptian uraeus-goddess, specifically identified as a raging "Eye," is said to have bound (*šnj*) the crown on Horus at the time of Beginning, thereby providing the King of the Gods with an invincible means of magical protection. A brief survey of the philology of *šnj* and related words helps to elucidate the celestial context of the symbolism in question. The word *šnṯ* signifies "snake" and has specific reference to a celestial serpent-dragon: "The serpent is in the sky...O *šnṯ* snake."[523] According to the Pyramid Texts (681), the *šnṯ*-dragon once imprisoned Horus. The *šnṯ*-snake is elsewhere described as an Ouroboros, with its tail in its mouth.[524] It is this serpent, apparently, which is described as encircling the king in an otherwise obscure Pyramid Text: "The King lies down in your coil, the King sits in your circle."[525] Similar ideas prevailed down through the three millennia of Egyptian civilization as indicated by an illustration from the Papyrus of Herytwebkhet, which shows a youthful Horus sitting within an Ouroboros-like enclosure (figure 3).

Figure 3

The image of the great god encircled by a serpentine enclosure recalls the so-called *shen*-bond, a popular Egyptian symbol otherwise known as the ring of sovereignty (see figure four).[526] Indeed Egyptologists have often remarked

---

[523] *PT* 444.
[524] *PT* 689. See now M. A. van der Sluijs & T. Peratt, "The Ourobóros as an Auroral Phenomenon," *Journal of Folklore Research* 46:1 (2009), pp. 3-41.
[525] *PT* 2289.
[526] S. Quirke, *Ancient Egyptian Religion* (London, 1992), p. 62.

upon the intimate relationship between the two symbols—i.e., the Ouroboros and *shen*-bond—although they have yet to succeed in offering a satisfactory explanation of the natural-scientific basis for the connection. Richard Wilkinson, for example, offered the following observation on the symbolism of the *shen*-bond:

"Being without beginning or end, the circle evokes the concept of eternity through its form, and its solar aspect is symbolized by the sun disk often depicted in the center of the shen sign. These ideas were probably the origin of this hieroglyph which is found in words connected with the verbal root shenu meaning 'encircle,' and which in its later elongated form became the cartouche which surrounded the Egyptian king's birth and throne names. Perhaps from this particular context the shen sign also took on the connotation of protection—as the device which excluded all inimical elements from the royal name. The shen may appear with both of these meanings—'eternity' and 'protection'—in Egyptian art…It [the shen hieroglyph] is also mirrored in the shape of the ouroboros, the serpent which bites its own tail."[527]

Figure 4

Egyptologists are agreed, moreover, that the *shen*-bond is intimately associated with archaic ideas of sovereignty and magical protection. Maria Betrò emphasized the bond's magical properties:

"This sign clearly represents a cord knotted at the ends and forming an oval; in the earliest period, however, it was definitely circular. Its meaning is clearly amuletic: it is the ring that magically isolates and protects that which is inside. Rings and ties—both positive and negative—are a recurring image in the practice and ideology of universal magic."[528]

The symbolism associated with the *shen*-bond is central to a proper understanding of Egyptian ideas of kingship and cosmic geography. Suffice it to note here that there is an intimate relationship between the raging uraeus-

---

[527] R. Wilkinson, *Reading Egyptian Art* (London, 1992), p. 193.
[528] M. Betrò, *Hieroglyphics* (New York, 1996), p. 195.

goddess and the enclosure known as the *shen*-bond, the latter of which houses the Egyptian pharaoh (as incarnation of the god Horus, the King of the Gods). And as if to emphasize the intimate connection between the encircling (*šnj*) uraeus-serpent and the *shen*-bond, in the sacred iconography celebrating the king's sovereignty the uraeus is depicted handing him the *šnw*-sign. Sally Johnson emphasized this particular role of the uraeus-goddess:

> *"She presents to the king's cartouche and Horus name the wȝs,* ↑*, scepter and šnw,* Q*, the signs for 'dominion' and 'infinity of the circuit of the sun', 'enclosure' or cartouche', thereby legitimizing his crown and sovereignty."*[529]

Venus and the Lead-Rope of Heaven

In ancient Egypt, as we have seen, the band encircling Horus was alternately conceptualized as a venom-spewing serpent, as a goddess of magic, as a crown, and as an encircling rope or knot (the *shen*-bond). That the band in question was celestial in nature is confirmed by the earliest texts.

Analogous ideas are attested in ancient Mesopotamia. Figure 5 shows Inanna/Venus holding the so-called "leadrope of heaven."[530] The formal resemblance between the leadrope (Akkadian *ṣerretu*) and the Egyptian *shen*-bond is evident at once. A common symbol of the king's terrestrial domain, Inanna's ring constituted a major insignia of kingship as indicated by numerous royal monuments.[531]

Figure 5

---

[529] S. Johnson, *The Cobra Goddess of Ancient Egypt* (London, 1990), p. 7.
[530] Adapted from an Old Babylonian plaque depicting Inanna/Ishtar in the Underworld. See figure 12 in J. Black et al, *The Literature of Ancient Sumer* (Oxford, 2004), p. 70.
[531] J. Klein, "The Coronation and Consecration of Šulgi in the Ekur (Šulgi G)," in H. Tadmor, M. Cogan & I. Eph'al (eds.) *Ah, Assyria...* (Jerusalem, 1991), p. 295. See also J. Black et al, *The Literature of Ancient Sumer* (Oxford, 2004), p. 98.

Similar traditions surround the Semitic goddess Ishtar/Venus, of whom it was said: "No one but she can [Hold the lead]rope of heaven."[532] The goddess herself is made to announce: "I grasp the leadrope of heaven in my hands."[533]

Literary texts also mention the planet Venus in connection with a celestial band. In the late Babylonian version of the hymn known as the *Exaltation of Ishtar*, the planet-goddess is described as holding the bond (*riksu*) of the sky. This passage has long troubled commentators. On the cosmic significance of the Ishtar's bond, Andrew George had this to say:

"A clue to the nature of the cosmological 'bond' comes from a second passage of the Exaltation of Ištar, in which the word ùz.sag, this time translated as riksu, denotes something that can be physically held... [Anu speaking to Ishtar] 'I, Anu, am the lord who directs (the heavens): take hold of their 'bond'! Holding their 'bond' is evidently the means by which Ištar is to control the heavens, now placed at her command. markasu is essentially a 'rope', particularly that by which a boat is secured to its mooring-post (tarkullu), and riksu, as also Sum. dur (=turru), has not only the vague notion of 'bond', but also the specific idea of a 'cord' that binds things together. It appears that the various parts of the Sumero-Babylonian universe were conceived as being linked or 'bonded' by one or more such cords or ropes."[534]

Absent from George's learned exposition is any discussion of how Ishtar's bond or rope is to be understood by reference to the familiar sky. Why would the planet Venus be associated with a cord-like bond encircling or "binding" the universe? Far from being a "vague notion of 'bond'," as per George, Ishtar's lead-rope is best understood as a perfectly visible and tangible celestial structure.

Wolfgang Heimpel, in fact, admitted that the bond of Ishtar/Venus had an astronomical significance. Thus, in a discussion of possible astronomical motifs in ancient Sumerian mythology, Heimpel observed:

"Many astral motifs are, in fact, attested:...Ištar's Elevation tells us that An married her on the urging of the gods and asked her to hold and rule the 'tie' (riksu) of the 'support' (pulukku) of the sky (III 13-36)."[535]

As for how we are to understand this celestial bond or "tie" from an astronomical standpoint, Heimpel offered nary a clue. From our vantage point, however, the answer is perfectly obvious: the rope-like bond of Inanna/Ishtar

---

[532] B. Foster, *Before the Muses* (Bethesda, 1993), p. 502.
[533] *Ibid.*, p. 901.
[534] A. George, *Babylonian Topographical Texts* (Leuven, 1992), pp. 261-262.
[535] W. Heimpel, "Mythologie, A. I," in E. Ebeling & B. Meissner eds., *Reallexikon der Assyriologie, Vol. 8* (Berlin, 1993-1997), p. 538.

has reference to the comet-like band that formerly spanned the heavens and comprised the enclosure of the gods.[536] It was this band that constituted the king's headband or crown and, as such, it can be viewed as an archetypal symbol of sovereignty. In the final analysis, then, the bond/tie of Inanna/Venus forms a precise structural parallel to the Egyptian *shen*-bond—the magical bond formed by the uraeus-serpent as it encircled Horus (Mars). In perfect harmony with this interpretation, the word *riksu* is elsewhere used of the magical spells employed by witches. Witness the following omen: "They (the witches) bind me with (their) plots (*irakkasani rik-si*)."[537]

Summary

Modern texts on Roman religion have precious little to say about the origins of the Latin goddess Venus. Recent investigations by Robert Schilling and other scholars have documented that Venus is intimately associated with archaic conceptions of magic—the goddess's name, as we have seen, has been interpreted to denote "charm" or "spell." As for how we are to understand the historical origins or phenomenological basis behind Venus's capacity for casting spells, or how this magical process might be reflected in the ancient cults of her presumed analogues from the ancient Near East—i.e., Aphrodite, Inanna, Ishtar, and Isis, among others—Schilling and his fellow Classicists have thus far remained silent.

By pursuing a comparative approach, we have established that the Latin goddess's propensity for casting magical spells likely has reference to her capacity for "binding." The same capacity is evident in Venus's intimate association with the royal crown: "She is connected with the *corona*, 'garland'... because the garland is a binder of the head and is itself, from *vinctura*, 'binding,' said *vieri* 'to be plaited,' that is, *vinciri*, to be bound."[538]

As a goddess much involved with magical spells and the "binding" on of the royal crown, the Latin Venus offers a close structural analogue to the Egyptian uraeus-goddess, the latter of whom was invoked as "Great-in-charms" and identified with the crown of kingship. As the Pyramid Texts state in no uncertain terms, the uraeus-goddess's capacity for magic and casting spells traces to her singular role in the "binding" or "encircling" (*šnj*) of Horus *in illo tempore*, whereupon she came to form the crown of kingship. Endlessly

---

[536] See the ground-breaking analysis in D. Talbott, *The Saturn Myth* (New York, 1980), pp. 145-171.
[537] *CAD*, "riksu," (Chicago, 1965-1989), p. 349. The passage in question is quoted from Maqlu IV 108.
[538] V 62, as quoted in Varro, *On the Latin Language* (Cambridge, 1967), pp. 59-61.

celebrated in Egyptian literature and ritual, this dramatic cosmogonic myth, in our view, has specific reference to cataclysmic events in which the uraeus-serpent—as the planet Venus—appeared to circumscribe Horus/Mars with a rope-like band.

The venom-spewing uraeus-goddess, in turn, forms a remarkable parallel to the venom-spewing Inanna, a planet-goddess who was intimately associated with the crown of sovereignty. And as the Egyptian uraeus-goddess came to be associated with a celestial band—the so-called *shen*-bond—so too did Inanna/Venus hold a celestial band in Sumerian lore. Inanna's *riksu*-tie, moreover, was specifically likened to a magical spell.

In the Sumerian language the word used to denote the fiery "venom" spewed forth by Inanna/Venus—uš$_{11}$—is explicitly compared to a magical "spell" or "charm," thereby paralleling the situation in the Latin language, wherein *venenum* signifies venom, poison, charm, and to cast a spell.[539] This circumstance, coupled with the complementary testimony presented earlier, leads us to the following conclusion: The fiery venom exuded by the dragon-like Inanna, like the fiery efflux exuded by the Egyptian uraeus-serpent, is inexplicable apart from the fiery efflux exuded by the comet-like Venus in prehistoric times. In short, the evidence gathered here suggests that the Latin cult of Venus reflects archaic conceptions regarding the planet Venus, albeit in vestigial form.

---

[539] J. Halloran, *Sumerian Lexicon* (Los Angeles, 2006), p. 305.

Conclusion

"If you have had your attention directed to the novelties of thought in your own lifetime, you will have observed that almost all really new ideas have a certain aspect of foolishness when they are first produced."[540]

In the present volume, the first in a multi-volume series, we have enumerated a few of the more compelling mythological themes involving the planet Venus. It has been documented that Venus was conceptualized as the prototypical female star, as an agent of war and disaster, as a star of lamentation, as a witch, and as a venom-spewing dragon. The fact that not one of these traditions finds a satisfactory explanation by reference to Venus's present appearance raises a number of important questions, not the least of which is what possessed the ancient skywatchers to describe the most prominent planet in such fashion?

In a brief overview of Inanna's cult in ancient Mesopotamia, we documented that the Sumerian planet-goddess was invoked as a raging warrior; as a terrifying agent of storm; and as a dragon raining fire from heaven. Early myths tell of her tragic love affair with the youthful Dumuzi, the latter described as a great warrior and brilliant star in heaven. A prominent rite in the New Year's celebration saw the king (as Dumuzi) simulating a "marriage" with Inanna/Venus—a marriage thought to legitimize the king's hold on the throne while ensuring fertility throughout the land. Turning to the New World, we discovered remarkably similar traditions amongst the Skidi Pawnee of the North American Plains: They, too, identified the planet Venus as the prototypical female power who, like Inanna, was represented as a formidable warrior. The planet-goddess eventually met her match in the Morning Star Mars who, as the most virile of warriors, first conquered and then impregnated her. It was this "marriage" between Venus and Morning Star that set Creation in motion. Most significant, perhaps, is the fact that the sacred rites designed to commemorate the *hieros gamos* between the two planets were believed to ensure fertility throughout the land. This remarkable correspondence in belief systems between two disparate cultures—hitherto unnoticed—supports our contention that the ancient traditions attached to Venus and Mars had an empirical basis and encode memorable astronomical events.

---

[540] Alfred North Whitehead, *Science and the Modern World*.

Analogous conceptions are to be found in ancient Greece: There, too, Aphrodite/Venus was intimately associated with marriage rites and identified as the paradigmatic bride in a sexual union with the Ares (Mars). Ares, like the Sumerian Dumuzi, was conceptualized as the paradigmatic bridegroom. Aphrodite, moreover, was described as the cosmic power who first coupled humans at the time of Creation, thereby mirroring similar traditions attached to the Skidi Venus. Recall again the Skidi prayer to Venus: "You shall be known as the Mother of all things; for through you all beings shall be created."[541]

How is it possible to understand such archaic traditions apart from a reference to astronomical phenomena? It is our contention that the Amerindian traditions serve to complement and clarify the Sumerian myth of Inanna and Dumuzi. The fact that the Skidi identified the planet Venus as the female agent in a prototypical "marriage of the stars" confirms the original celestial context of the Sumerian sacred marriage rite—a fact consistently overlooked by entire generations of scholars on ancient myth and religion. Equally important is the fact that the Skidi myth allows us to deduce the celestial prototype for Dumuzi: Inanna's youthful paramour is to be identified with the planet Mars. As a warring star and consort of the planet Venus, Dumuzi offers a close analogue to the Skidi Morning Star Mars.

Earlier we posed the following question: Why would Sumerian kings seek (or expect) legitimization for their sovereignty through a symbolic marriage with the planet Venus, the latter personified as Inanna? In ancient Mesopotamia, as around the globe, a king could only acquire legitimacy by emulating or reenacting an archetypal prototype—in this case by emulating the behavior of Dumuzi/Mars *in illo tempore*, at which time the planetary hero united in marriage with Inanna/Venus and, as a result, gained sovereignty and all its glories (empowerment, beautification, apotheosis, etc.). By engaging in a ritualized sexual union with the planet Venus, Sumerian kings were evidently operating under the widespread belief that by carefully simulating the prototypical cosmogonic events in question a sympathetic magical result could be effected—i.e., it would lead to *their* empowerment and deification while promoting fertility throughout the land (just as a sudden increase in fertility and vegetation accompanied Dumuzi's marriage with Inanna/Venus).

What was true in ancient Mesopotamia was true around the globe: It was the planet Venus that embodied sovereignty and, in the guise of the Mother Goddess, it was she who invested the king with his regal powers and crown of glory. This belief-system prevailed in ancient Persia, as we have documented, where Anahita/Venus represented sovereignty and conferred

---

[541] George Dorsey, *Traditions of the Skidi Pawnee* (Boston, 1904), p. 3.

charisma or power on kings and heroes. The charisma conferred by Anahita/Venus, in this sense, is indistinguishable from the charisma conferred on Ares by Aphrodite.[542] The charm associated with the Latin goddess Venus belongs here as well and originated in dramatic astronomical events (charm and charisma are recognized cognates).

Analogous beliefs are attested in ancient Egypt, where Isis/Venus was explicitly identified with the throne and thought to imbue the king—as Horus—with his regal powers. It is the "embrace" of the Mother Goddess that results in the "beautification" of Horus. Witness Jan Assmann's summary of Isis as "kingmaker":

"As mother of the Horus child, Isis...is not only the great healing goddess, but also the bestower of legitimate kingship. Her milk not only heals illness, it makes the child a king, it 'creates,' as the Egyptian terminology puts it, his 'beauty.' Isis is the 'kingmaker' par excellence... In Egypt, the legitimate salvation-bringing king was not the 'anointed one,' but the 'suckled one.' Many temple reliefs, particularly from the New Kingdom, represent him in this role, in the arms and at the breast of the mother-goddess Isis."[543]

If recounting the central events of Creation comprised a primary function of ancient myth, it stands to reason that sacred traditions around the globe will tell of the scintillating union of Venus and Mars. That said, the ravages of time and disjunctions due to euhemerization and/or historicization have taken their toll with the predictable result that this greatest of all romance tales is not always recognizable as an explosive encounter of planetary powers. Who among us would discern a coupling of planets in Homer's telling of Aphrodite's liaison with Ares?

The modifying effects of "humanization" and creative embellishment are aptly illustrated by the Star Woman myth. Ostensibly recounting an amorous encounter between a Star Woman and a homely mortal, a prominent motif finds the latter being transfigured as a result of their sexual liaison—usually he is described as beautified or otherwise rejuvenated or empowered. The Greek myth of Aphrodite and Phaon is paradigmatic in this regard, but very similar traditions are to be found around the globe.

In South America, as we have documented, Star Woman is expressly identified with the planet Venus. This datum offers compelling support for our thesis that early traditions surrounding Aphrodite originated in the dramatic

---

[542] B. Breitenberger, *Aphrodite and Eros* (New York, 2007), p. 114.
[543] J. Assmann, "Death and Initiation in the Funerary Religion of Ancient Egypt," in W. Simpson ed., *Religion and Philosophy in Ancient Egypt* (New Haven, 1989), p. 134.

recent history of the planet Venus. The Greek goddess's intimate association with sovereignty,[544] marriage-rites,[545] and sacred gardens points to the same conclusion.[546]

The Star Woman myth also serves to elucidate the celestial context of the sacred marriage between Inanna and Dumuzi. Thus it is that Star Woman's imbuing of her mortal paramour with "radiance" or "beauty" forms a close structural analogue to the curious episode in BM 96739 whereby Dumuzi is transfigured and empowered by Inanna's splendor, the latter described as an outpouring of luminous "power" or strength. Properly understood, the Star Woman myth—like the marriage of Inanna and Dumuzi—commemorates an extraordinary conjunction of planets whereby the planet Mars was enveloped by the catastrophically charged corona of Venus.

Most important, perhaps, is the fact that our survey of Venus-lore has revealed that catastrophic imagery lies at its very core. Whether it is the image of Inanna/Venus as a serpent-dragon raining fire from the sky, or Hathor's rampage as the raging "Eye," the destructive behavior ascribed to the planet is emphasized in myth after myth.

Particularly telling are those traditions describing Venus as having fallen from heaven as a meteor or comet-like fireball. Recall the Sumerian tradition recounting the fall of Lamashtu-Inanna; the Phoenician ritual celebrating Astarte's fall as a fiery meteor; and the Shipibo account from South America wherein Venus is described as falling from heaven as a meteor amidst great noise and tumult. The fact that Aphrodite's oldest fetish at Paphos was a black meteorite likely reflects analogous conceptions.

Equally compelling and catastrophically inspired are those mythological traditions that describe the great goddess as suffering a dramatic metamorphosis in appearance whereupon she appears as a witch-like hag with wildly disheveled hair. Prominent examples of this widespread motif include Lamashtu/Inanna, al-'Uzza, Holda, and Kali. As the examples provided by Lamashtu and al-'Uzza confirm, there is an indissoluble relation between the disheveled goddess and the planet Venus. Hence we understand why the ancient Babylonian skywatchers described the planet Venus as a "witch-star," an appellation that will never be

---

[544] As witnessed by the epithet *Basileia*. Of the latter goddess, Karl Kerényi, *Goddesses of Sun and Moon* (Dallas, 1979), p. 44 observed: "Through marriage to this most beautiful maiden, one would receive lordship over the whole world. The word '*Basileia*,' differently accented, would commonly mean kingdom."

[545] Rachel Rosenzweig, *Worshipping Aphrodite* (Ann Arbor, 2004), p. 21 writes: "The artistic evidence makes it abundantly clear that Aphrodite is the *sine qua non* of wedding ritual."

[546] *Ibid.*, p. 30: "Before the goddess even reached the Greek mainland she was honored as a vegetal deity on Paphos as *Hierokepia*—the equivalent of the Athenian title *en Kepois*."

explained by reference to that planet's current appearance. So, too, it is not by accident that the Inca described Venus as the planet with disheveled hair.

Archetypal myths like the *hieros gamos* between Venus and Mars did not arise *de novo* or from some dark recesses of the Collective Unconscious. Nor, for that matter, do they commemorate the familiar and perennially peaceful interactions of these two planets. Rather, there is a wealth of evidence that such myths commemorate truly extraordinary encounters with awe-inspiring planetary powers—an experience otherwise known as a *mysterium tremendum et fascinosum*. As Thorkild Jacobsen and other scholars have argued, this otherworldly experience is fundamental to all religion: "A confrontation with a 'Wholly Other' outside of normal experience and indescribable in its terms; terrifying, ranging from sheer demonic dread through awe to sublime majesty; and fascinating, with irresistible attraction, demanding unconditional allegiance."[547]

Such, in our opinion, was the essence of the Earthlings' encounter with Venus and her pockmarked paramour. This all-too-close and terrifying encounter with the "gods" *in illo tempore* left an indelible mark on the human Psyche that endures to this very day, not all of which is conscious in nature. If we aspire to understand the fundamental message of ancient myth—not to mention the origin and stubborn persistence of religious belief systems—it is essential to recognize the archetypal motifs and recurring thematic patterns as the direct biological (behavioral/emotional) response to witnessed *historical* events. Then, and only then, will a reality-based reconstruction of Earth history and religion be possible.

---

[547] T. Jacobsen, *The Treasures of Darkness* (New Haven, 1976), p. 3.

Appendices

Why Study Myth?

"What is to be made of this mass of nonsense? How can all this have a meaning, a motivation, a function, or at least a structure? The question of whether myths have an authentic content can never be put in positive terms."[548]

Why should anyone care about mythological traditions recounting a sexual liaison between Venus and Mars? Ancient myth represents, as it were, a mnemonic "fossil" reflecting the intellectual history of mankind. For untold millennia the narration and memorization of such stories remained the primary means of transmitting valued information about the history of the world and the most treasured beliefs of early man. If we are to gain a better understanding of the origins of religion, folklore, and human culture itself, it stands to reason that we would do well to study the content and message of ancient myth wherein such matters form a central concern.

Especially significant from the standpoint of modern science is the information encoded in ancient myth regarding the recent history of the solar system. Indeed, it is our contention that the eyewitness testimony of ancient man—as recorded in sacred traditions and rock art the world over—offers a surprisingly detailed and trustworthy guide for reconstructing that history.

An old adage proclaims that the devil is in the details. This could well serve as the motto for developing a science of mythology, for in comparing and analyzing myths it is the recurring and often incongruous details of structure and language which frequently betray affinities and significant informational content. In this sense the scientific study of mythology resembles nothing

---

[548] P. Veyne, *Did the Greeks Believe in their Myths* (Chicago, 1988), p. 2.

so much as comparative anatomy—the science of identifying and clarifying structural and functional affinities between different animals. To the expert eye, an isolated molar together with the odd fibula or femur, will reveal unequivocal signs of form and function, thereby aiding in the reconstruction of the entire animal and its probable way of life and phylogenetic history.

In the biological sciences in general, and in comparative anatomy in particular, it is the overarching theme of evolution that provides the historical framework and theoretical rationale for understanding how and why particular animals came to look the way they do. Embryonic whales occasionally display vestigial hip-sockets and hind limbs for one reason and one reason alone—because they formerly descended from land mammals.[549] Yet even a relatively informed and diligent observer of whales in the natural world could easily overlook this compelling evidence for evolution inasmuch as modern whales no longer display external hind limbs. Appearances notwithstanding, the anatomist's scalpel readily reveals the diminutive hip-sockets and limbs just the same. And as a testament to historical origins, one could hardly ask for more telltale evidence than the whale's vestigial structures.

And so it is with countless structures of ancient myth and ritual, wherein a seemingly anomalous mytheme or formulaic phrase stands out as if a "throwback" to a distant age, often in apparent contempt for narrative context or a common sense understanding of the natural world. As a case in point, consider the myriad of mythological traditions which describe a warrior-goddess or witch wandering the world with disheveled hair—this despite the fact that such creatures *do not exist* in the real world. As bizarre as these traditions appear at first sight, comparative analysis will show that such motif-structures as the goddess's disheveled hair and storm-raising are every bit as telling as the whale's vestigial hind limbs in revealing the historical origins—and celestial prototype—of the warrior-goddess/witch archetype.

As the theory of evolution provides the overarching model whereby we can understand innumerable biological structures and behavioral patterns, it also allows biologists to reconstruct a prototypical "Ur-form" from which a particular species or phylum descended—this despite the fact that such a form has yet to be discovered in the extant paleontological record. Thus it is that "whales with legs" were postulated long before they were ever documented in the fossil record.

In the study of ancient myth, in contrast to the situation prevailing in the biological sciences, a unifying theory has yet to be advanced which can provide

---

[549] P. Gingerich, "Paleobiological Perspectives on Mesonychia, Archaeoceti, and the Origin of Whales," in J. Thewissen ed., *The Emergence of Whales* (New York, 1998), pp. 423-449.

the necessary theoretical model and historical framework whereby the world's vast corpus of mythological structures might be classified or understood. It is our opinion that the planetary-catastrophism model defended by Talbott and myself represents just such a unifying concept and, as such, constitutes an important breakthrough and methodological first step towards the desired goal of developing a science of mythology.[550] And much as a thorough familiarity with the facts of anatomy and paleontology equips the biologist with the requisite knowledge to recognize the whale's fundamental affinity with other mammals, both living and long extinct, so, too, does a comprehensive knowledge of ancient lore enable the comparativist to discern underlying thematic patterns and structures shared between disparate sacred traditions—even those which, on the surface, would appear to bear little resemblance to each other.

In the study of ancient myth, as in evolutionary biology, there is simply no substitute for the comparative method. No matter how learned the researcher, and no matter how rigorous his reasoning, it is virtually impossible to understand the origins of a particular myth from the vantage point of one culture alone. The reason for this is perfectly obvious although it seems to have gone virtually unrecognized to date. Given the fact that myths have been told and retold for countless millennia, it stands to reason that each culture's mythological corpus has been profoundly impacted by "mutations"; i.e., secondary accretions and embellishments which modify or otherwise distort its original message. If so, it follows that theoretical analyses based on the traditions of one particular culture alone are likely to be compromised or undermined by the "noise" emanating from these secondary mutations and local developments such as humanization and/or historicization. Yet if analysis is focused primarily on those mythological structures (mythemes) that are found around the globe, one can be relatively confident that these extraneous mutations can be recognized and controlled for analytical purposes. A comparative analysis thus lessens the likelihood of arriving at false conclusions due to the loss of basic structures or the faulty transmission of the original myth. Most important, perhaps, is the fact that the comparative analysis of analogous mythological structures—much like the comparative analysis of individual bones collected from different animals widely separated in time and space—allows for the reconstruction of a prototypical Ur-myth from which the respective "daughter myths" have descended. This remains true even in those cases wherein a myth's original historical context may have been lost to a particular culture or region.

---

[550] See now M. van der Sluijs, *On the Origin of Myths in Catastrophic Experience, Vol. 1* (Vancouver, 2019).

Given its apparent antiquity and wide range of distribution, the Star Woman myth constitutes an exemplary case study illustrating the theoretical rationale of comparative mythology. If the original myth can be represented by the alphabet—wherein the 26 letters constitute the myth's entire set of mythemes or structural components—the respective extant versions can be symbolized as follows: ABC1, DEF2, ACF3, MOP4, etc. While no one version preserves the entire mythological repertoire, a comparative analysis of the attested versions will allow for the isolation and recovery of the various structural components and this, in turn, will point us towards the myth's original form (i.e., A-Z). Through comparative analysis it will eventually become evident which structural components constitute mutations and/or arbitrary developments (represented here by the respective numerals), as when a particular tribe in South America mistakenly identifies Star Woman as Jupiter rather than Venus.[551] Upon isolating the various structures and thematic patterns shared by different narratives, and culling out the secondary accretions, it should be possible to reconstruct the original myth *and its natural historical origins.*

Ideally, a scientific analysis of the Star Woman theme will seek to understand the various local variants by reference to the myth's original structure and natural-historical context. The secondary accretions, in turn, will generally be ignored as "noise," much as paleontologists ignore genetic anomalies or injuries in their attempt to reconstruct the prototypical "whale." By proceeding in this manner, it should be possible to place the respective local variants in their original mytho-historical context even though that context has been otherwise largely obscured.

Such an approach stands in marked contrast to that practiced by other schools of mythological exegesis, which all too often becomes bogged down in the analysis of secondary developments and therefore fails to recognize the archetypal mythological structures and thematic patterns. Especially guilty in this regard is Claude Lévi-Strauss, whose arcane method of analysis considers each particular version of a myth to be equally valid and informative.[552] This is a classic example of mistaking the individual trees and attendant undergrowth for the proverbial forest, and thus it is little wonder that the pioneering French anthropologist ended up advancing such absurdities as the following dictum: "myths get thought in man unbeknownst to him."[553]

---

[551] C. Nimuendajú, "Šerenté Tales," *Journal of American Folklore* 57 (1944), p. 184.

[552] C. Lévi-Strauss, "The Structural Study of Myth," in T. Sebeok, *Myth: A Symposium* (London, 1955), p. 92.

[553] C. Lévi-Strauss, *Myth and Meaning* (New York, 1995), p. 3.

The primary goal of comparative mythological analysis is, or ought to be, the reconstruction of the historical determinants behind the genesis of a particular mythological structure—the so-called mythemes. In this way—and in this way only—in our opinion, will it be possible to decipher the information encoded in ancient mythological traditions. If the thesis developed here has any validity, it follows that the world's vast corpus of myth and ritual will contain innumerable structures pointing to planetary catastrophe, many of which survive as mere vestiges and isolated reminiscences. Like the rudimentary hind limbs of embryonic whales, such mnemonic "throwbacks" represent compelling evidence of ideational structures that originated in a specific natural-historical context and once served a specific functional purpose, now barely recognizable and devoid of function. Inanna's "king-making" marriage with Dumuzi is but one of thousands of such vestigial mythological structures.

Suns and Planets in Prehistoric Rock Art

> *"We may consider the ancients' perception of the stars in the sky a pure metaphor but for many of them it had apparently empirical reality, it was simply what they saw."*[554]
>
> *"Astronomy always has been and still is a science that relies on the use of past observations. Unlike most sciences, astronomy can never be truly experimental: astronomers can only observe the astronomical phenomena that present themselves...Perhaps uniquely in the sciences, astronomers, therefore, are forced to rely upon empirical data collected by their predecessors."*[555]

The discovery in 1879 of spectacular paintings in the caves of Altamira (northern Spain) was initially met with disbelief and ridicule, so radical was the idea that artwork of such skill and beauty could have been created by people living in the Stone Age. It was only after the discovery of similar finds in France, Portugal and elsewhere in Europe that the scientific community was forced to accept the reality of Paleolithic rock art. In the years since that discovery it has been documented that rock art is present upon all inhabited continents and spans a period of time measured in millennia (the paintings of Altamira and Lascaux are typically dated to circa 15-35,000 BCE).[556]

During the Paleolithic Age, rock art was devoted primarily to the naturalistic depiction of various forms of wildlife, presumably objects of the hunt or associated in some way with rites of sympathetic magic.[557] Paintings of horses and wisent, the great bison that once roamed the steppes of Europe, are especially common, although mammoths, woolly rhinoceroses, and other long extinct fauna also appear. Depictions of the sun, moon, or other familiar celestial objects such as the Milky Way are evidently absent from these earliest artworks. Indeed, so far as we can determine, out of the many thousands of paintings from the Paleolithic Age there is not a single clearly discernible image of the sun or moon.[558]

---

[554] G. Selz, "The Tablet with 'Heavenly Writing', or How to Become a Star," in A. Panaino ed., *Non licet stare caelestibus* (Udine, 2014), p. 55.

[555] J. Steele, *Ancient Astronomical Observations and the Study of the Moon's Motion (1691-1757)* (London, 2012), pp. 3-4.

[556] A. Willcox, *The Rock Art of Africa* (Kent, 1984), pp. 1-5.

[557] H. Breuil & H. Obermaier, *The Cave of Altamira* (Madrid, 1935), p. 12; F. Windels, *The Lascaux Cave Paintings* (London, 1949), p. 137.

[558] G. Curtis, *The Cave Painters* (New York, 2006), p. 17.

It was during the subsequent Neolithic Age, apparently, that artists first began recording their impressions of celestial phenomena through paintings and petroglyphs (incised images in rock). Not unlike fossilized bones, which provide paleontologists with a vast database upon which to reconstruct the precise structure and relative abundance of prehistoric fauna, rock art represents an objective record of mankind's enduring interest in the stars and, as such, offers a rich compendium upon which to test reconstructions of the solar system's recent history.

Among the most common petroglyphs are those interpreted as early images of the sun. Included here are relatively simple images featuring a circular disc from which "rays" emanate in all directions (see figure 1).[559]

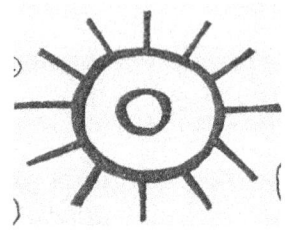

Figure 1

Other images, however, are more difficult to reconcile with the appearance of the present solar orb. Consider the image in figure 2, depicting what would appear to be a circular disc with a smaller orb set within its center.[560] Of this particular image, Emmanuel Anati remarked:

"This kind of symbolic representation of the sun is common to many primitive societies and ancient civilizations. It occurs in the ancient Near East, in the Far East, as well as in Europe and elsewhere."[561]

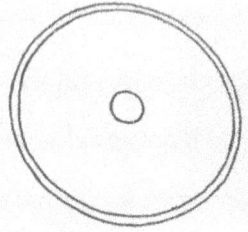

Figure 2

---

[559] This image is adapted from figure 329 in R. Heizer & C. Clewlow, *Prehistoric Rock Art of California* (Ramona, 1973).

[560] This image is adapted from E. Anati, *Camonica Valley* (New York, 1961), p. 162.

[561] *Ibid.*, p. 47.

Even more difficult to square with the present sun is the image represented in figure 3, showing a flower-like object set within the center of a circular disc (the fact that the image in question occurs in the general context of other "solar" images confirms its celestial provenance).[562] Although less common than the image represented in figure 2, this particular petroglyph also has precise parallels around the globe.[563]

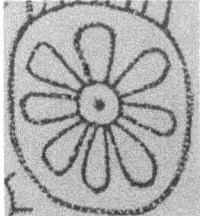

Figure 3

Consider also the image illustrated in figure 4.[564] How is it possible to explain the peculiar wheel-like "spokes" radiating outwards across the disc by reference to the familiar solar orb?[565] Most perplexing, perhaps, is the fact that such petroglyphs are not only found around the globe they occur in Neolithic contexts and thus predate by several millennia the actual invention of spoked wheels: "The wheel-like symbols cannot *be* wheels since passage grave art appeared some millennia before the spoked wheel came into being in these regions."[566]

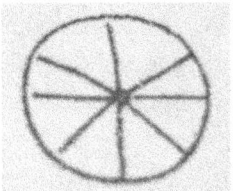

Figure 4

---

[562] This image, taken from Cairn T of Lough Crew, forms figure 235 in E. Twohig, *op. cit.*
[563] See figure 438 of B. Teissier, *Ancient Near Eastern Cyllinder Seals* (Berkeley, 1984), p. 225. For a parallel from the New World, see R. Heizer & C. Clewlow, *op. cit.*, figure 85.
[564] This image is adapted from Cairn U at Loughcrew. See M. Brennan, *The Stones of Time* (Rochester, 1994), p. 160. The very same image appears in abundance at Val Camonica. For an analogue from Africa, see figure 8:4 from A. Willcox, *The Rock Art of Africa* (Kent, 1984). For an analogue from the New World, see figure 47w from R. Heizer & C. Clewlow, *Prehistoric Rock Art of California* (Ramona, 1973).
[565] Of these wheel-like forms, E. Anati, *Camonica Valley* (New York, 1961), p. 163 observes: "A number of hypotheses have been advanced in an attempt to explain them, but none is truly satisfactory."
[566] M. Green, *The Sun-Gods of Ancient Europe* (London, 1991), p. 25.

Other petroglyphs depict what appear to be ladder-like appendages projecting from the so-called solar image (see figure 5).[567] Here, too, we are dealing with a petroglyph of widespread distribution, one typically interpreted as the sun with "rays."

Figure 5

Although the aforementioned "sun-images" occur in a wide variety of artistic contexts and mediums, it is not uncommon to find them associated with scenes of apparent worship and ritual. In some engravings, for example, people are depicted "offering salutations" to the sun with upraised arms. In Camonica Valley (northern Italy), one of the richest and most thoroughly explored petroglyph sites in the world, Anati reported: "The carvings of the first period are limited to the depiction of one person praying, facing the sun—which is drawn as a disc with a dot in the center."[568] Such scenes, coupled together with abundant evidence suggesting that the sun featured prominently in ancient pantheons, have led scholars to assume that the petroglyphs in question served some sort of religious purpose for the Stone Age artists and their communities.[569]

Reviewing the aforementioned images, it is evident at once that they have no obvious counterpart in the present skies. How, then, are we to interpret them? As mere random doodlings? As the product of drug-induced hallucinations of shamans and cave men? For Miranda Green, author of *The Sun-Gods of Ancient Europe*, the artworks in question are a product of creative imagination: "The pictures do not describe reality."[570] The same author elsewhere adds the curious observation:

---

[567] Adapted from figure 253 in R. Heitzer & C. Clewlow, *Prehistoric Rock Art of California*, Vol. 2 (Ramona, 1973). For examples of "suns" with pillars, see E. Anati, *op. cit.*, p. 162. See also the discussion in E. Cochrane, "Ladder to Heaven," *Aeon* 6:5 (2004), pp. 55-76.
[568] E. Anati, *op. cit.*, p. 47.
[569] *Ibid.*, p. 230. See also M. West, *Indo-European Poetry and Myth* (Oxford, 2007), p. 198.
[570] M. Green, *The Sun-Gods of Ancient Europe* (1991), p. 76.

"Man did not simply look at the sun and copy what he saw to the best of his ability. He went further and interpreted and superimposed new images of the sun which were not based entirely on his visual perception."[571]

Such interpretations are entirely baseless and wrongheaded, in our opinion. The mere fact that virtually identical images are attested around the globe argues strongly against Green's position. From our vantage point there is no more reason to question the realism of these so-called solar petroglyphs than the prehistoric paintings of horses and bison at Altamira and Lascaux. No one would ever think of dismissing the latter artworks as a product of fantasy, hallucination, or creative imagination. Why, then, should the anomalous solar petroglyphs be dismissed out of hand as nonrepresentational simply because they fail to conform with the present skies? For us it makes more sense to interpret both types of artworks as representational in nature—as relatively faithful attempts to accurately convey the lived experience and natural world of the prehistoric cave artists. Yet once you open your mind to this possibility, it becomes very difficult to put the genie back in the bottle for the simple willingness to take the ancient artists' testimony at face value naturally raises the question as to how to explain the specific imagery of the different celestial artworks.

In the pages to follow we will be discussing a seemingly bewildering array of different "solar" images, all of which are anomalous by reference to the prevailing skies. It is our position that each of these images has its own story to tell. In order to advance the scientific study of archaeoastronomy—not to mention comparative mythology itself—the way forward must commence by first journeying backwards in time, to the very origins of civilization in ancient Mesopotamia and Egypt. A survey of the earliest artworks and cosmogonic traditions of these two civilizations, it turns out, will reveal one glaring anomaly after another, each pointing to a radically different solar system, hitherto overlooked by all scholars.

---

[571] *Ibid.*, p. 33. Elsewhere, p. 20, the same author writes: "The artist, even of a veristic image, is unable to transcribe exactly what he sees."

Solar Imagery in Ancient Mesopotamia

Mesopotamia is rightly renowned as the birthplace of astronomy. From time immemorial, Babylonian skywatchers scanned the sky for signs of impending disaster and evil omen, eventually collecting their observations and omens together in a series of texts known as the *Enuma Anu Enlil*, compiled during the second millennium BCE. It was these particular omen texts that Ptolemy consulted when attempting to place Greek astronomical practice on a rational scientific foundation (Babylonian astronomical practices were early on diffused abroad and influenced the development of the sciences in China, India, and Arabia).[572] Given the unrivaled antiquity and broad scope of their observations, the Mesopotamian astronomical records and traditions remain indispensable for the modern scholar attempting to piece together humankind's earliest memories and conceptions associated with the Sun, stars, and planets.

Complimenting the astronomical texts are the so-called cylinder seals, engravings cut into various types of stone that originally served as signs of property ownership. Deriving from earlier stamp seals, cylinder seals first appeared in the mid-fourth millennium BCE and are generally regarded as among the "high points of Mesopotamian craftmanship."[573] Early cylinder seals commonly depict images believed to represent familiar celestial bodies. Figure one, for example, is universally held to depict the sun.[574] Why Mesopotamian artists would select this particular image to serve this function remains unclear, as the current solar orb does not display a central dot. That said, the very same image is ubiquitous around the globe, occurring in both historic and prehistoric

---

[572] K. Stevens, *Between Greece and Babylonia* (Cambridge, 2019), pp. 42-46.
[573] C. Fischer, "Twilight of the Sun-God," *Iraq* 64 (2002), p. 125. See also O. Topcuoglu, "Iconography of Protoliterate Seals," in C. Woods ed., *Visible Language* (Chicago, 2015), p. 29.
[574] Adapted from figure II:8 in L. Werr, *Studies in the Chronology and Regional Style of Old Babylonian Cylinder Seals* (Malibu, 1988). See also P. Amiet, *La glyptique mésopotamienne archaique* (Paris, 1961), figure 1641.

contexts.⁵⁷⁵ In the earliest pictographic scripts in Egypt and China, moreover, this very sign served as an ideogram for "sun."⁵⁷⁶

Figure 1

Similar questions arise with regard to the image depicted in figure 2, which shows a solar disc with an eight-rayed star inscribed in its center.⁵⁷⁷ In Mesopotamian iconography, the eight-rayed star is known to represent the planet Venus.⁵⁷⁸ How, then, are we to explain the fact that early cylinder seals seemingly depict the Venus-star as superimposed on the "sun"-disc and enclosed within a "lunar" crescent? Dominique Collon—a leading authority on Mesopotamian cylinder seals—offered the following opinion on this strange state of affairs:

"From Ur III times onwards, however, the crescent is also often combined with a disc inscribed with a star which is placed within it (star-disc and crescent...). This could either be explained as different phases of the moon or, more likely, is a shorthand for the principal celestial bodies, sun (and star?) and moon."⁵⁷⁹

---

⁵⁷⁵ E. Cochrane, "Suns and Planets in Neolithic Rock Art," in *Martian Metamorphoses* (Ames, 1997), pp. 194-214.
⁵⁷⁶ J. Norman, *Chinese* (New York, 1988), p. 61.
⁵⁷⁷ Adapted from figure III:5 in L. Werr, *Studies in the Chronology and Regional Style of Old Babylonian Cylinder Seals* (Malibu, 1988).
⁵⁷⁸ F. Rochberg, "Heaven and Earth," in S. Noegel & J. Walker eds., *Prayer, Magic, and the Stars in the Ancient and Late Antique World* (University Park, 2003), pp. 174-176 writes: "The association of the heavenly bodies with certain deities seems to go back to the very beginnings of Mesopotamian civilization and persists as well to the end. Astral emblems, such as the lunar crescent (Akk. uškaru) for Sin, the eight-pointed star for Ištar, and the solar disc (Akk. šamšatu) for Šamaš, are a regular feature of Mesopotamian iconography throughout its history. These divine symbols can be traced on cylinder seals as early as the Early Dynastic period and as late as the Neo-Babylonian."
⁵⁷⁹ D. Collon, "Mond," *RA* 8 (Berlin, 1993-1997), p. 357.

Figure 2

At this point the open-minded researcher must confront certain basic questions of common sense and logic: How likely is it that the most celebrated astronomers of the ancient world would have insisted upon depicting the three most prominent celestial bodies in astronomically impossible positions? Was it sheer perversity alone that inspired the ancient artists to reproduce these particular images again and again?

Gennadij Kurtik is one of the few scholars to even acknowledge the jarring anomaly presented by these archaic sun-discs with the Venus-star depicted in the center. Yet his explanation of the hybrid celestial symbols amounts to little more than a wild guess:

"Since the period of the Akkade Dynasty (XXIV-XXII centuries BC), ... the astral symbol of Inanna (an eight-pointed star) was frequently found inscribed in a circle. Why? The answer is probably in some poetic texts of the New-Sumerian period (XXII-XXI centuries BC); for example, in the hymn by Iddin-Dagan devoted to Inanna her shining in the night is compared with the light of day or the Sun...the attribute of being solar is transferred to Inanna, therefore the solar disk is becoming her symbol."[580]

Kurtik and Collon are seemingly unaware of the fact that artworks depicting "stars" set within the center of a "sun"-disc are to be found around the globe. Witness the African image depicted in figure 3.[581] Here, too, the mere fact that analogous images are commonplace on other continents as well suffices to rule out the manifestly ad hoc explanations offered by Collon and Kurtik, which would interpret the anomalous "solar" images as the product of metaphorical musings unique to ancient Mesopotamia.

---

[580] "The Identification of Inanna with the Planet Venus," *Astronomical and Astrophysical Transactions* 17 (1999), p. 508.
[581] Adapted from figure 21 in G. Williams, *African Designs* (New York, 1971).

Figure 3

In addition to the eight-pointed star, the planet-goddess Inanna-Ishtar could also be denoted by several other symbols. A rosette, for example, commonly adorns artistic scenes and objects deemed sacred to the great goddess. As to the antiquity of this symbol, Elizabeth van Buren noted: "From the earliest times the rosette was a symbol of the goddess Innin-Ishtar."[582]

Significantly, early examples of the rosette feature eight "petals" or arms extending out from a central dot and thus bear a striking resemblance to an eight-pointed star (see figure 4).[583] Van Buren argues that the two symbols are likely cognate: "The eight-pointed star of Istar, frequently illustrated on monuments of the second and third millennia, was an adapted form of the archaic rosette as may be clearly seen from the star carved at the top of a kudurru from Susa."[584]

Figure 4

A close variation on this symbol appears in early Mesopotamian artworks supposedly depicting the Sun (see figure 5), thereby confirming its celestial provenance.[585]

---

[582] E. van Buren, *Symbols of the Gods in Mesopotamian Art* (Rome, 1945), p. 84.
[583] Adapted from figure 8 in U. Moortgart-Correns, "Die Rosette—ein Schriftzeichen?," in *Altorientalische Forschungen* 21 (1994), p. 369. See also E. van Buren, "The Rosette in Mesopotamian Art," *Zeitschrift für Assyriologie XI* (1939), pp. 99-107.
[584] E. van Buren, "The Rosette in Mesopotamian Art," *ZA* 45 (1939), p. 105.
[585] Adapted from figure 264 in W. Ward, The Seal Cylinders of Western Asia (Washington D. C., 1910), p. 91. See also BM 22963 in D. Collon, *First Impressions* (London, 1987), p. 42.

Figure 5

Our discussion of Venus-symbolism has direct relevance to the question posed earlier: What is the significance of the eight-pointed star set in the middle of the Shamash-disc? Given the intimate association between the star and Venus, what is the relation between the eight-pointed star of the Shamash disc and the Venus star? Does the convergence of iconography imply that artistic license prevailed among ancient astronomers? Or does it perhaps commemorate some hitherto unrecognized relationship between Venus and the ancient sun-god?

As we have documented elsewhere, a survey of the Mesopotamian cylinder seals will reveal dozens of anomalous "suns," none of which bears any resemblance to the current solar orb.[586] In numerous representations of the ancient sun, for example, the eight-pointed star presents a wheel-like appearance (see figure 6).[587]

Figure 6

---

[586] E. Cochrane, "Anomalies in Ancient Descriptions of the Sun-God," *Chronology & Catastrophism Review* (2016), pp. 3-12.

[587] Adapted from figure V:9 in L. Werr, *Studies in the Chronology and Regional Style of Old Babylonian Cylinder Seals* (Malibu, 1988).

An equally common symbol on Mesopotamian cylinder seals finds the central star being depicted as a four-rayed, diamond-like form (figure 7 depicts a so-called "sun"-image from the Akkadian period, thought to originate from around 2300 BCE).[588] Here, too, however, virtually identical images will be found around the globe, thereby supporting the conclusion that these peculiar images depict a conspicuous celestial object or constellation, albeit one unknown to modern astronomical science. Figure 8 features a similar image from an Egyptian bowl dating to the predynastic period (circa mid-4th millennium BCE).[589] Figure 9 depicts a stellar form common throughout the American Southwest.[590]

Figure 7   Figure 8

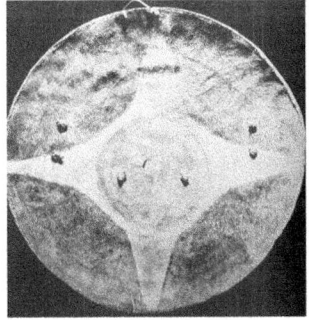

Figure 9

---

[588] Adapted from figure IV:4 in L. Werr, *Studies in the Chronology and Regional Style of Old Babylonian Cylinder Seals* (Malibu, 1988). See also J. Black & A. Green, *Gods, Demons and Symbols of Ancient Mesopotamia* (London, 1992), p. 168: "The disc with four-pointed star and three radiating wavy lines between each of the points occurs from the Akkadian down to the New Babylonian Period. It almost invariably stands as a symbol of the sun god Šamaš (Utu)."

[589] Eva Wilson, *Ancient Egyptian Designs* (London, 1986), figure 8.

[590] Adapted from Barton Wright, *Pueblo Shields* (Flagstaff, 1976), p. 50. It will be noted that the same basic image denoted "sun" in the Nahuatl pictographic script. See M. León-Portilla & E. Shorris, *In the Language of Kings* (New York, 2001), p. 4.

The similarity between these three images is evident at once, as is the fact that they bear no discernible resemblance to the appearance of the present sun (the solar orb can never appear inside a lunar crescent, for example). How, then, are we to account for their origin? The fact that such images are to be found around the globe and share a number of unique structures in common suggests that we have to do here with relatively faithful depictions of a former "sun" or constellation and *not* with any arcane "shorthand" or metaphor peculiar to the Mesopotamian mindset. Yet historians of ancient art never consider the possibility that these iconic symbols might faithfully represent the appearance of a lost "sun," one unknown to modern science.

Bibliography

H. Alexander, "North American," in L. Gray ed., *The Mythology of All Races, Vol. 10* (Boston, 1917).
S. Allan, *The Shape of the Turtle* (Albany, 1991).
J. Allen, *Genesis in Egypt* (New Haven, 1988).
J. Allen, "The Cosmology of the Pyramid Texts," in J. Allen et al eds., *Religion and Philosophy in Ancient Egypt* (New Haven, 1989), pp. 1-28.
J. Allen, *The Ancient Egyptian Pyramid Texts* (Atlanta, 2005).
T. Allen, *Horus in the Pyramid Texts* (Chicago, 1916).
B. Alster, *Dumuzi's Dream* (Copenhagen, 1972).
W. Andrae, *Die Ionische Säule* (Berlin, 1933), pp. 20-67.
C. Andrews, *Amulets of Ancient Egypt* (Austin, 1994).
R. Anthes, "Egyptian Theology in the Third Millennium B.C.," *JNES* 18 (1959), pp. 169-212.
R. Anthes, "Horus als Sirius in den Pyramidentexten," *ZÄS* 102 (1975), pp. 1-10.
J. Assmann, "Horizont," *LÄ* III (Berlin, 1977), cols. 3-7.
J. Assmann, *The Mind of Egypt* (Cambridge, 2002).
J. Assmann, *The Search for God in Ancient Egypt* (Ithaca, 2001).
H. Balz-Cochois, *Inanna* (Gütersloh, 1992).
H. Baumann, *Schöpfung und Urzeit des Menschen im Mythus der afrikanischen Völker* (Berlin, 1936).
L. Bean, "Menil (Moon): Symbolic Representations of a Cahuilla Woman," in R. Williamson & C. Farrer eds., *Earth and Sky* (Albuquerque, 1992), pp. 162-183.
P. Beaulieu, *The Pantheon of Uruk during the Neo-Babylonian Period* (Leiden, 2003).
H. Behrens, *Die Ninegalla-Hymne* (Stuttgart, 1998).
M. Betrò, *Hieroglyphics* (New York, 1996).

J. Black et al, *The Electronic Text Corpus of Sumerian Literature* (http://www-etcsl.orient.ox.ac.uk/) (Oxford, 1998).

J. Black et al, *The Literature of Ancient Sumer* (Oxford, 2004).

C. Blinkenberg, *The Thunderweapon in Religion and Folklore* (Cambridge, 1911).

F. Boas, *Indianische Sagen von der Nord-Pacifischen Küste Amerikas* (Berlin, 1895).

F. Boas, *Tsimshian Mythology* (Washington D. C., 1916).

J. F. Borghouts, "The Evil Eye of Apophis," *Journal of Egyptian Archaeology* 59 (1973), pp. 114-150.

J. F. Borghouts, "Magic," in *LÄ* (Wiesbaden, 1972-1992), cols. 1137-1151.

P. Breutz, "Sotho-Tswana Celestial Concepts," in *Ethnological and Linguistic Studies in Honour of N.J. van Warmelo* (Pretoria, 1969), pp. 199-210.

D. Brinton, *The Myths of the New World* (New York, 1968).

D. Brown, *Mesopotamian Planetary Astronomy-Astrology* (Groningen, 2000).

F. Brown & S. Driver & C. Briggs, *A Hebrew and English Lexikon* (Oxford, 1951).

B. Brundage, *The Fifth Sun* (Austin, 1979).

F. Bruschweiler, *Inanna la déesse triomphante et vaincue dans la cosmologie sumérienne* (Leuven, 1988),

E. Budge, *The Egyptian Book of the Dead* (London, 1901).

E. Budge, *The Gods of the Egyptians, Vol. I* (New York, 1969).

E. Budge, *Amulets and Talismans* (New York, 1968).

E. van Buren, "The Sacred Marriage in Early Times in Mesopotamia," *Orientalia* 13 (1944), pp. 1-72.

E. Van Buren, *Symbols of the Gods in Mesopotamian Art* (Rome, 1945).

W. Burkert, *Greek Religion* (Cambridge, 1985).

E. Butterworth, *The Tree at the Navel of the Earth* (Berlin, 1970).

L. Cagni, *The Poem of Erra* (Malibu, 1977).

H. Cairns, "Aboriginal sky-mapping," in C. Ruggles ed., *Archaeoastronomy in the 1990's* (Leicestershire, 1993), pp. 136-152.

D. Cardona, "The Sun of Night," *Kronos* 3:1 (1977), pp. 31-38; "The Mystery of the Pleiades," *Kronos* 3:4 (1978), pp. 24-44.

J. Carlson, "Transformations of the Mesoamerican Venus Turtle Carapace War Shield: A Study in Ethnoastronomy," in V. del Chamberlain et al eds., *Songs From the Sky* (Leicester, 2005), pp. 99-122.

R. Carmack, *Quichean Civilization* (Berkeley, 1973).

A. Carnoy, "Iranian Views of Origins," *Journal of American Oriental Society* 36 (1916), pp. 300-320.

E. C. L. During Caspers, "The Gate-Post in Mesopotamian Art...," *Jaarbericht ex Oriente Lux* (1971/2), pp. 211-227.

P. Cate, "The Hittite Storm God: his Role and his Rule According to Hittite Cuneiform Sources," in D. Meijer, *Natural Phenomena* (Amsterdam, 1992), pp. 83-148.

P. Cattermole, *Venus: The Geological Story* (Baltimore, 1994).

R. Chadwick, "Identifying Comets and Meteors in Celestial Observation Literature," in H. Galter ed., *Die Rolle der Astronomie in den Kulturen Mesopotamiens* (Graz, 1993), pp. 161-184.

V. Del Chamberlain, *When Stars Came Down to Earth* (College Park, 1982).

V. Del Chamberlain & P. Schaafsma, "Origin and Meaning of Navaho Star Ceilings," in V. del Chamberlain et al eds., *Songs From the Sky* (Leicester, 2005), pp. 80-98.

P. Chemery, "Meteorological Beings," M. Eliade ed., *The Encyclopedia of Religion, Vol. 9* (New York, 1987), pp. 487-492.

M. de Civrieux, *Watunna: An Orinoco Creation Cycle* (San Francisco, 1980).

R. Rundle Clark, *Myth and Symbol in Ancient Egypt* (London, 1959),

P. A. Clarke, "The Aboriginal Cosmic Landscape of Southern South Australia," *Records*

of the South Australian Museum 29:2 (1997), pp. 125-145.

B. Cobo, *Inca Religion and Customs* (Austin, 1990).

E. Cochrane & D. Talbott, "When Venus was a Comet," *Kronos* 12:1 (1987), pp. 2-24.

E. Cochrane, "Mars Gods of the New World," *Aeon* 4:1 (1995), pp. 47-63.

E. Cochrane, *The Many Faces of Venus* (Ames, 2001).

E. Cochrane, *Starf*cker* (Ames, 2006).

M. Cohen, *Sumerian Hymnology* (Cincinnati, 1981).

A. B. Cook, *Zeus, Vol. 2* (New York, 1965).

A. Coomaraswamy, "The Symbolism of the Dome," in R. Lipsey ed., *Coomaraswamy: Selected Papers* (Princeton, 1977), pp. 415-458.

A. Coomaraswamy, "Svayamatrnna: Janua Coeli," in R. Lipsey ed., *Coomaraswamy: Selected Papers* (Princeton, 1977), pp. 465-520.

A. Coomaraswamy, *Symbolism of Indian Architecture* (Jaipur, 1983).

F. Cumont, *Astrology and Religion Among the Greeks and Romans* (New York, 1960).

E. M. Curr, *The Australian Race, Vol. II* (Melbourne, 1886).

S. Dalley, *Myths from Mesopotamia* (Oxford, 1991).

U. Dall'Olmo, "Latin Terminology Relating to Aurorae, Comets, Meteors, and Novae," *Journal for the History of Astronomy* 11 (1980), pp. 10-27.

H. Davidson, *Gods and Myths of Northern Europe* (Baltimore, 1964).

J. van Dijk, "Inanna raubt den 'grossen Himmel': Ein Mythos," In S. Maul ed., *Festschrift für Rykle Borger* (Groningen, 1994), pp. 9-38.

R. Dixon, "Oceanic Mythology," in L. Gray ed., *The Mythology of All Races* (Boston, 1916).

G. Dorsey & J. Murie, "Notes on Skidi Pawnee Society," *Field Museum of Natural History* 27:2 (1940), pp. 73-119.

G. Dumézil, *Archaic Roman Religion, Vol. 1* (Baltimore, 1996).

G. Dumézil, *The Destiny of the King* (Chicago, 1973).

W. Eilers, *Sinn und Herkunft der Planetennamen* (München, 1976).

M. Eliade, *Patterns in Comparative Religion* (New York, 1958).

M. Eliade, *Shamanism* (Princeton, 1964).

M. Eliade, *Rites and Symbols of Initiation* (New York, 1958).

A. Erman, *Hymnen an das Diadem der Pharaohen* (Berlin, 1911).

C. Eyre, *The Cannibal Hymn* (Liverpool, 2002).

L. Farnell, *The Cults of the Greek States, Vol. II* (New Rochelle, 1977).

C. Faraone, *Talismans and Trojan Horses* (New York, 1992).

R. Faulkner, "The Bremner-Rhind Papyrus— IV," *Journal of Egyptian Archaeology* 24 (1938), pp. 41-58.

R. Faulkner, "The King and the Star-Religion in the Pyramid Texts," *JNES* 25 (1966), pp. 153-161.

R. Faulkner, *The Ancient Egyptian Pyramid Texts* (Oxford, 1969).

R. Faulkner, *The Ancient Egyptian Coffin Texts* (Warminster, 1973-1978).

R. Faulkner, *The Egyptian Book of the Dead* (San Francisco, 1994).

W. Fauth, "Ištar als Löwengöttin und die löwenköpfige Lamaštu," *Die Welt des Orients* 12 (1981), pp. 21-36.

L. Fison, *Tales from Old Fiji* (London, 1894).

E. Florescano, *The Myth of Quetzalcoatl* (Baltimore, 1999).

B. Foster, "Ea and Saltu," in M. de Jong Ellis ed., *Essays on the Ancient Near East in Memory of J.J. Finkelstein* (New Haven, 1977), pp. 79-84.

B. Foster, *From Distant Days* (Bethesda, 1995).

B. Foster, *Before the Muses: An Anthology of Akkadian Literature* (Bethesda, 2005).

H. Frankfort, *Kingship and the Gods* (Chicago, 1948).

H. Frankfort, *The Art and Architecture of the Ancient Orient* (1954).

D. Frayne, "Notes on The Sacred Marriage Rite," *Bibliotheca Orientalis* 42:1/2 (1985), cols. 5-22.

J. Frazer, *Myths of the Origin of Fire* (London, 1930).

J. Frazer, "Phaethon and the Sun," in *Apollodorus: The Library Vol. II* (Cambridge, 1963), pp. 388-394.

J. Frazer, *Folklore in the Old Testament* (New York, 1988).

J. Frazer, *Adonis, Attis, Osiris* (New Hyde Park, 1961).

D. Freidel, L. Schele, & J. Parker, *Maya Cosmos* (New York, 1993).

W. Gaerte, "Kosmische Vorstellungen im Bilde prähistorischer Zeit: Erdberg, Himmelsberg, Erdnabel und Weltenströme," *Anthropos* 9 (1914), pp. 956-979.

I. Gamer-Wallert, "Heiliger Baum," *LÄ, Vol. I* (Berlin, 1977), cols. 655-660.

A. Gardiner, "The Personal Name of King Serpent," *Journal of Egyptian Archaeology* 44 (1958), pp. 38-39.

T. Gaster, *Myth, Legend, and Custom in the Old Testament* (New York, 1969).

A. George, *Babylonian Topographical Texts* (Leuven, 1992).

J. Glassner, *The Invention of Cuneiform* (Baltimore, 2003).

M. Godelier, "Myth and History," *New Left Review* 69 (1971), pp. 93-112.

D. Goetz & S. Morley, *Popol Vuh* (Norman, 1972).

J. Gonda, *Epithets in the RgVeda* (S-Gravenhage, 1959).

J. Gonda, *Aspects of Early Visnuism* (Delhi, 1969).

P. Gössmann, *Planetarium Babylonicum* (Rome, 1950).

A. K. Grayson, "Assyria," in J. Boardman et al eds., *The Assyrian and Babylonian Empires and other States of the Near East, from the Eighth to the Sixth Centuries B.C.* (Cambridge, 1991), pp. 103-141.

A. Green, "Ancient Mesopotamian Religious Iconography," in J. Sasson ed., *Civilizations of the Ancient Near East, Vols. 3/4* (Farmington Hills, 1995), pp. 1837-1856.

M. Green, *Symbol and Image in Celtic Religious Art* (London, 1989).

M. Green, *The Sun-Gods of Ancient Europe* (London, 1991).

M. Green & H. Nissen, *Zeichenliste der archaischen Texte aus Uruk* (Berlin, 1987).

T. Griffin-Pierce, "Ethnoastronomy in Navaho Sandpaintings of the Heavens," *Archaeoastronomy* 9 (1986), pp. 62-69.

R. Griffith, *The Rig Veda* (New York, 1992).

J. Grimm, *Teutonic Mythology* (Gloucester, 1976).

G. Grinnell, "Some Early Cheyenne Tales," *The Journal of American Folk-Lore* 20 (1907), pp. 169-194.

G. Grinnell, "Some Early Cheyenne Tales: II," *The Journal of American Folk-Lore* 21 (1908), pp. 269-320.

D. Grinspoon, *Venus Revealed* (New York, 1997).

O. Gurney, "The Sultantepe Tablets," *Anatolian Studies* 10 (1960), pp. 105-131.

H. Haeberlin, "Mythology of Puget Sound," *Journal of American Folklore* 37 (1924), pp. 371-438.

B. Haile, *Starlore Among the Navaho* (Sante Fe, 1977).

W. Hallo & J. van Dijk, *The Exaltation of Inanna* (New Haven, 1968).

J. Halloran, *Sumerian Lexicon* (Los Angeles, 2006).

R. Hannig, *Ägyptisches Wörterbuch I* (Mainz, 2003).

J. Harrison, *Epilegomena to the Study of Greek Religion and Themis* (New Hyde Park, 1962).

J. Hayes, *A Manual of Sumerian Grammar and Texts* (Malibu, 2000).

W. Heimpel, "A Catalog of Near Eastern Venus Deities," *Syro-Mesopotamian Studies* 4 (1982), pp. 59-72.

W. Heimpel, "Mythologie, A. I," *Reallexikon der Assyriologie, Vol. 8* (Berlin, 1993-1997), pp. 537-564.

W. Heimpel, "The Sun at Night and the Doors of Heaven," *Journal of Cuneiform Studies* 28:2 (1986), pp. 127-151.

R. Heitzer & C. Clewlow, *Prehistoric Rock Art of California, Vol. 2* (Ramona, 1973).

A. Hiltebeitel, "Draupadī's Hair," in M. Biardeau ed., *Autour de la déesse hindoue* (Paris, 1981), pp. 179-214.

A. Hiltebeitel, *The Ritual of Battle* (Ithaca, 1976).

U. Holmberg, "Finno-Ugric, Siberian Mythology," in L. Gray ed., *The Mythology of All Races, Vol. IV* (Boston, 1927).

E. Hornung, "Dat," *Lexikon Ägyptologie, Vol. I* (Berlin, 1977), col. 994.

E. Hornung, *Conceptions of God in Ancient Egypt* (Ithaca, 1982).

E. Hornung, *Idea Into Image* (Princeton, 1992).

E. Hornung, "Ancient Egyptian Religious Iconography," in J. Sasson ed., *Civilizations of the Ancient Near East, Vols. 3/4* (Farmington Hills, 1995), pp. 1711-1730.

W. Horowitz, *Mesopotamian Cosmic Geography* (Winona Lake, 1998).

T. Hoskinson, "Saguaro Wine, Ground Figures, and Power Mountains," in R. Williamson & C. Farrer, *Earth and Sky* (Albuquerque, 1992), pp. 131-161.

C. Houtman, "Queen of Heaven," in K. van der Toorn et al eds., *Dictionary of Deities and Demons in the Bible* (Leiden, 1995), cols. 1278-1283.

B. Hruška, "Das spätbabylonische Lehrgedicht 'Inanna's Erhöhung'," *Archiv Orientální* 37 (1969), p. 473-521.

Å. Hultkrantz, *The Religions of the American Indians* (Berkeley, 1967).

J. Isaacs, *Australian Dreaming, 40,000 years of Australian History* (Sydney, 1980).
W. G. Ivens, *Melanesians of the South-east Solomon Islands* (London, 1927).
T. Jacobsen, *The Treasures of Darkness* (New Haven, 1976).
T. Jacobsen, *The Harps That Once...* (New Haven, 1987).
E. O. James, *Creation and Cosmology* (Leiden, 1969).
A. Jeremias, *Handbuch der altorientalischen Geisteskultur* (Leipzig, 1913).
A. Jeremias, "Schamasch," in W. Roscher ed., *Ausfuhrliches Lexikon der griechischen und römischen Mythologie* (Hildesheim, 1965), cols. 533-558.
R. Jewell, *Pacific Designs* (London, 1998).
D. Johnson, *Night Skies of Aboriginal Australia* (Sydney, 1998).
S. Johnson, *The Cobra Goddess of Ancient Egypt* (London, 1990).
T. Judt, *Reappraisals* (New York, 2008).
D. Katz, *The Image of the Netherworld in the Sumerian Sources* (Bethesda, 2003).
Y. Ke, *Dragons and Dynasties* (Singapore, 1991).
E. Keber, *Codex Telleriano Remensis* (Austin, 1995).
O. Keel & C. Uehlinger, *Gods, Goddesses, and Images of God* (Minneapolis, 1998).
A. Keith, "Indian Mythology," in L. Grey ed., *The Mythology of All Races, Vol. 6* (Boston, 1917).
C. Kerényi, *The Gods of the Greeks* (London, 1982).
J. King, "A Southeastern Native American Tradition: The Ofo Calendar and Related Sky Lore," *Archaeoastronomy* 14:1 (1999), pp. 109-135.
D. Kinsley, *The Sword and the Flute* (Berkeley, 1975).
J. Klein, "The Coronation and Consecration of Šulgi in the Ekur (Šulgi G)," in H. Tadmor, M. Cogan & I. Ephœal (eds.) *Ah, Assyria...* (Jerusalem, 1991), pp. 292-313.
A. Kötz, *Über die astronomischen Kenntnisse der Naturvölker Australiens und der Südsee* (Leipzig, 1911).
S. Kramrisch, *The Presence of Siva* (Princeton, 1981).
R. Krauss, *Astronomische Konzepte und Jenseitsvorstellungen in den Pyramidentexten* (Wiesbaden, 1997).
R. Krauss, "The Eye of Horus and the Planet Venus: Astronomical and Mythological References," in J. Steele & A. Imhausen eds., *Under One Sky* (Münster, 2002), pp. 193-208.
E. Krupp, *Beyond the Blue Horizon* (Oxford, 1991).
E. Krupp, "Phases of Venus," *Griffith Observer* 56:12 (1992), pp. 2-18.

F. Kuiper, *Ancient Indian Cosmogony* (Leuven, 1983).

R. Kutscher, "The Cult of Dumuzi/Tammuz," in J. Klein ed., *Bar-Ilan Studies in Assyriology* (New York, 1990), pp. 29-44.

W. Lamb, "Star Lore in the Yucatec Maya Dictionaries," in *Archaeoastronomy in Pre-Columbian America* (Lubbock, 1975), pp. 233-248.

W. Lambert, "Studies in Nergal," *Bibliotheca Orientalis* 30:5/6 (1973), pp. 355-363.

W. Lambert, "Lugal-IGI.DU-anna," *Reallexikon der Assyriologie* 7 (Berlin, 1983), p. 142.

W. Lambert, "The History of the muš-ḫuš in Ancient Mesopotamia," in U. Seidl ed., *l'animal, l'homme, le dieu dans le proche-orient ancien* (Leuven, 1985), pp. 87-94.

S. Langdon, "Semitic Mythology," in L. Gray ed., *The Mythology of All Races* (New York, 1964).

P. Lapinkivi, *The Sumerian Sacred Marriage* (Helsinki, 2004).

R. Lehmann-Nitsche, "Mitologia sudamericana," *Revista del Museo de La Plata* 27 (1923/1925), pp. 267-285.

G. Leick, *A Dictionary of Ancient Near Eastern Mythology* (London, 1991).

B. Lesko, *The Great Goddesses of Ancient Egypt* (Norman, 1999).

L. Lesko, "Ancient Egyptian Cosmogonies and Cosmology," in B. Shafer ed., *Religion in Ancient Egypt* (Ithaca, 1991), pp. 88-122.

C. Lévi-Strauss, *The Raw and the Cooked* (Chicago, 1969).

H. Liddell & R. Scott, *A Greek-English Lexicon* (New York, 1872).

R. Linton, "The Sacrifice to Morning Star by the Skidi Pawnee," *Leaflet Field Museum of Natural History, Department of Anthropology* 6 (1923), pp. 1-18.

R. Linton, "The Origin of the Skidi Pawnee Sacrifice to the Morning Star," *American Anthropologist* 28 (1928), pp. 457-466.

B. MacLachlan, *The Age of Grace: Charis in Early Greek Poetry* (Princeton, 1993).

J. Major, "Myth and Origins of Chinese Science," *Journal of Chinese Philosophy* 5 (1978), pp. 1-20.

J. Major, *Heaven and Earth in Early Han Thought* (Buffalo, 1993).

T. Mann, "Freud and the Future," *Daedalus* 88 (1959), pp. 374-378.

S. Markel, *Origins of the Indian Planetary Deities* (Lewiston, 1995).

T. McCleary, *The Stars We Know: Crow Indian Astronomy and Lifeways* (Prospect Heights, 1997).

B. Meador, *Inanna: Lady of the Largest Heart* (Austin, 2000).

S. Mercer, *The Pyramid Texts in Translation and Commentary* (New York, 1952).

A. Métraux, "Myths and Tales of the Matako Indians," *Ethnological Studies* 9 (1939), pp. 1-127.

A. Métraux, *Myths of the Toba and Pilaga Indians of the Gran Chaco* (Philadelphia, 1946).

T. Mettinger, *The Riddle of Resurrection* (Stockholm, 2001).

S. Milbrath, *Star Gods of the Maya* (Norman, 1999).

D. Miller, *Stars of the First People* (Boulder, 1997).

J. Monroe & R. Williamson, *They Dance in the Sky* (Boston, 1987).

O. Montelius, "The Sun God's Axe and Thor's Hammer," *Folklore* 21 (1910), pp. 60-78.

C. Mountford, *Art, Myth and Symbolism* (Melbourne, 1956).

C. Mountford, *Arnhem Land: Art, Myth and Symbolism* (Melbourne, 1968).

J. Murie, "Ceremonies of the Pawnee," *Smithsonian Contributions to Anthropology* 27 (Washington D. C., 1981), pp. 1-182.

G. Murray, *The Collected Plays of Euripides* (London, 1954).

G. Nagy, *The Ancient Greek Hero* (Cambridge, 2013).

M. Naylor, *Authentic Indian Designs* (New York, 1975).

O. Neugebauer & R. Parker, *Egyptian Astronomical Texts, Vol. 3* (London, 1960).

E. Neumann, *The Great Mother* (Princeton, 1972).

C. Nimuendaju, "The Šerenté," *Publication of the Frederick Webb Hodge Anniversary Publication Fund, Vol. 4* (Los Angeles, 1942).

R. Onians, *The Origins of European Thought* (New York, 1973).

A. Oppenheim, "Man and Nature in Mesopotamian Civilization," in C. Gillispie ed., *Dictionary of Scientific Biography, Vol. 15* (New York, 1978), pp. 634-666.

W. Otto, *Dionysus* (Bloomington, 1965).

E. Parsons, *Pueblo Indian Religion* (Chicago, 1939).

A. Peratt, "Characteristics for the Occurrence of a High-Current, Z-Pinch Aurora as Recorded in Antiquity," *IEEE Transactions on Plasma Science* 31:6 (2003), pp. 1192-1214.

C. Di Peso, "Prehistory: O'otam," in A. Ortiz ed., *Handbook of North American Indians: Southwest, Vol. 3* (Washington, 1979), pp. 91-100.

R. Pinxten & I. Van Dooren, "Navajo Earth and Sky," in R. Williamson & C. Farrer, *Earth and Sky* (Albuquerque, 1992), pp. 101-109.

E. Polomé, "Some Thoughts on the Methodology of Comparative Religion, with Special Focus on Indo-European," in E. Polomé ed., *Essays in Memory of Karl Kerényi* (Washington, D.C., 1984), pp. 9-27.

E. Polomé, "Germanic Religion: An Overview," in *Essays on Germanic Religion* (Washington, D.C., 1989), pp. 68-138.

N. Postgate, T. Wang & T. Wilkinson, "The evidence for early writing...," *Antiquity* 69 (1995), pp. 459-480.

B. Pritzker, *A Native American Encyclopedia* (Oxford, 2000).

J. Puhvel, *Comparative Mythology* (Baltimore, 1987).

R. Redfield & A. Rojas, *Chan Kom—A Maya Village* (Chicago, 1964).

G. Reichel-Dolmatoff, *Amazonian Cosmos* (Chicago, 1971).

D. Reisman, *Two New-Sumerian Hymns* (1970), p. 166. Note: This was a dissertation presented to the University of Pennsylvania.

D. Reisman, "Iddin-Dagan's Sacred Marriage Hymn," *Journal of Cuneiform Studies* 25 (1973), pp. 185-202.

L. Reitzammer, *The Athenian Adonia in Context* (Madison, 2016).

F. Reynolds, "Unpropitious Titles of Mars in Mesopotamian Scholarly Tradition," in J. Prosecky ed., *Intellectual Life of the Ancient Near East* (Prague, 1998), pp. 347-358.

A. Risser, "Seven Zuni Folk Tales," *El Palacio* 48 (1941), pp. 215-230.

R. Ritner, *The Mechanics of Ancient Egyptian Magical Practice* (Chicago, 1993).

J. B. Rives et al, "Venus," in H. Cancik & H. Schneider eds., *Der Neue Pauly* 12:2 (Stuttgart, 2000), cols. 17-20.

A. Roberts, *Hathor Rising* (Devon, 1995).

P. Roe, *The Cosmic Zygote* (New Brunswick, 1982).

P. Roe, "Mythic Substitution and the Stars..." in V. Del Chamberlain et al eds., *Songs From the Sky* (Sussex, 2005), pp. 193-228.

W. Römer, "Beitrage zum Lexikon des Sumerischen," *Bibliotheca Orientalis* XXXII: 5/6 (1975), pp. 145-162.

W. Roscher, *Die Gorgonen und Verwandtes* (Leipzig, 1879).

L. Rose, "A Critique of Peter Huber," in L. Greenberg & W. Sizemore, eds., *Velikovsky and Establishment Science* (Glassboro, 1977), pp. 102-112.

W. Roth, "An Inquiry into the Animism and Folk-Lore of Guiana Indians," *Bureau of American Ethnology* 30 (1915), pp. 254ff.

F. Russell, *The Pima Indians* (Washington, 1908).

C. Sagan, *Comet* (New York, 1985).

B. de Sahagún, *Florentine Codex, Book 7* (Sante Fe, 1950-1970).

J. Sando, "Jimez Pueblo," in A. Ortiz ed., *Handbook of North American Indians: Southwest, Vol. 3* (Washington, 1979), pp. 418-429.

G. de Santillana & H. von Dechend, *Hamlet's Mill* (Boston, 1969).

P. Schaafsma, *Warrior, Shield, and Star* (Sante Fe, 2000).

J. Scheid, "Venus," in S. Hornblower & A. Spawforth eds., *The Oxford Classical Dictionary* (Oxford, 1996), p. 1587.

W. K. Schenkel, "Horus," *LÄ* III (Berlin, 1977), cols. 14-25.

R. Schilling, *La Religion romaine de Vénus* (Paris, 1954).

W. Schwartz, *Indogermanischer Volksglaube* (Berlin, 1885).

Y. Sefati, *Love Songs in Sumerian Literature* (Jerusalem, 1998).

G. Selz, "Five Divine Ladies," *NIN* 1 (2000), pp. 29-62.

K. Sethe, Übersetzung und Kommentar zu den altägyptischen Pyramidentexten, Vol. 1 (Wiesbaden, 1962).

H. von Sicard, "Karanga Stars," *NADA* 19 (1943), pp. 42-65.

Å. Sjöberg & E. Bergmann, *The Collection of the Sumerian Temple Hymns* (Locust Valley, 1969).

Å. Sjöberg, "in-nin šà-gur$_4$-ra. A Hymn to the Goddess Inanna...," *Zeitschrift für Assyriologie* 65 (1976), pp. 161-253.

Å. Sjöberg, *Der Mondgott Nanna-Suen* (Stockholm, 1980).

Å. Sjöberg, "In the Beginning," in T. Abusch ed., *Riches Hidden in Secret Places* (Winona Lake, 2002), pp. 229-247.

R. Smyth, *The Aborigines of Victoria, Vol. 1* (London, 1878).

W. Staudacher, *Die Trennung von Himmel und Erde* (Darmstadt, 1968).

P. Steinkeller, "Inanna's Archaic Symbol," in J. Braun et al eds., *Written on Clay and Stone* (Warsaw, 1998), pp. 87-99.

P. Steinkeller, "On Rulers, Priests and Sacred Marriage," in K. Watanabe ed., *Priests and Officials in the Ancient Near East* (Tokyo, 1996), pp. 103-137.

M. Stol, "The Moon as seen by the Babylonians," in D. J. Meijer ed., *Natural Phenomena, Their Meaning, Depiction, and Description in the Ancient Near East* (Amsterdam, 1992), pp. 245-278.

V. Straizys & L. Klimka, "Cosmology of the Ancient Balts," *Journal of the History of Astronomy* 28 (1997), pp. S57-S81.

B. Stross, "Venus and Sirius: Some Unexpected Similarities," *Kronos* XII:1 (1987), pp. 25-42.

P. Sturrock, *Plasma Physics* (Cambridge, 1994).

W. Sullivan, *The Secret of the Incas* (New York, 1996).

K. Szarzynska, "Cult of the Goddess Inana in Archaic Uruk," in *Sumerica* (Warsaw, 1997), pp. 141-153.

K. Szarzynska, "Some of the Oldest Cult Symbols in Archaic Uruk," *Jaarbericht ex Oriente Lux* 30 (1987-88), pp. 3-21.

K. Szarzynska, "Offerings for the goddess Inana," *Revue d'assyriologie et d'archéologie orientale* 87 (1993), pp. 7-26.

K. Szarzynska, *Sumerica* (Warsaw, 1997).
D. Talbott, *The Saturn Myth* (New York, 1980).
D. Talbott, "Servant of the Sun God," *Aeon* 2:1 (1989), pp. 37-52.
D. Talbott, *Symbols of An Alien Sky* (Portland, 1997).
D. Talbott & W. Thornhill, *Thunderbolts of the Gods* (Portland, 2005).
K. Tallqvist, *Akkadische Götterepitheta* (Helsinki, 1938).
B. Tedlock, "Zuni Sacred Theater," *American Indian Quarterly* 7 (1983), pp. 93-110.
B. Tedlock, "Maya Astronomy: What We Know and How We Know It," *Archaeoastronomy* 18 (1999), pp. 39-58.
D. Tedlock, *Popol Vuh* (New York, 1985).
S. Thompson, *Tales of the North American Indians* (Bloomington, 1966).
M. Thomsen, *The Sumerian Language* (Copenhagen, 1984).
W. Tindale, "The Legend of Waijungari...," *Records of the South Australian Museum* 5:3 (1935), pp. 261-274.
D. Tunbridge, *Flinders Ranges Dreaming* (Canberra, 1988).
M. Varro, *On the Latin Language* (Cambridge, 1967).
I. Velikovsky, *Worlds in Collision* (New York, 1950).
I. Velikovsky, *Mankind in Amnesia* (New York, 1982).
C. Villacorta & J. Villacorta, *The Dresden Codex* (Walnut Creek, 1992).
J. de Vries, *Altnordische etymologisches Wörterbuch* (Leiden, 1977).
H. Wagenvoort, "The Origin of the Goddess Venus," in *Pietas* (Leiden, 1980), pp. 166-196.
B. Warner, "Traditional Astronomical Knowledge in Africa," in C. Walker ed., *Astronomy Before the Telescope* (London, 1996), pp. 304-317.
E. von Weiher, *Der babylonische Gott Nergal* (Berlin, 1971).
G. Weltfish, *The Lost Universe* (New York, 1965).
R. Wertime & A. Schuster, "Written in the Stars: Celestial Origin of Maya Creation Myth," *Archaeology* 46:4 (July/August, 1993), pp. 26-35.
L. Werr, *Studies in the Chronology and Regional Style of Old Babylonian Cylinder Seals* (Malibu, 1988).
M. West, *Indo-European Poetry and Myth* (Oxford, 2007).
J. Westenholz, *Legends of the Kings of Akkade* (Winona Lake, 1997).
Joan Westenholz, "Goddesses of the Ancient Near East 3000-1000 BC," in L. Goodison & C. Morris eds., *Ancient Goddesses* (London, 1998), pp. 63-82.
F. Wiggermann, "Mischwesen. A," *Reallexikon der Assyriologie* 8 (Berlin, 1993-1997), pp. 222-245.

F. Wiggermann, "*muš∆uššu*," *Reallexikon der Assyriologie* 8 (Berlin, 1993-1997), pp. 455-462.

F. Wiggerman, "Nergal," *Reallexikon der Assyriologie* 9 (1999), pp. 215-226.

J. Wilbert & K. Simoneau, *Folk Literature of the Chamacoco Indians* (Los Angeles, 1987).

J. Wilbert & K. Simoneau, *Folk Literature of the Chorote Indians* (Los Angeles, 1985).

J. Wilbert & K. Simoneau, *Folk Literature of the Makka Indians* (Los Angeles, 1991).

J. Wilbert & K. Simoneau, *Folk Literature of the Mataco Indians* (Los Angeles, 1982).

J. Wilbert & K. Simoneau, *Folk Literature of the Mocovi Indians* (Los Angeles, 1988).

J. Wilbert & K. Simoneau, *Folk Literature of the Nivalké Indians* (Los Angeles, 1987).

J. Wilbert & K. Simoneau, *Folk Literature of the Sikuani Indians* (Los Angeles, 1992).

J. Wilbert & K. Simoneau, *Folk Literature of the Toba Indians, Vol. 2* (Los Angeles, 1989).

C. Wilcke, "Inanna/Ishtar," *Reallexikon der Assyriologie* 5 (Berlin, 1976-1980), pp. 74-87.

R. Wilkinson, *Reading Egyptian Art* (London, 1992).

T. Wilkinson, *Early Dynastic Egypt* (London, 2001).

R. Williamson, *Living the Sky* (Norman, 1984).

R. Williamson, *Religious Beliefs and Cosmic Beliefs of Central Polynesia, Vol. 1* (Cambridge, 1933).

D. Wolkstein & S. Kramer, *Inanna* (New York, 1983).

E. Worms, *Australian Aboriginal Religions* (Richmond, 1986).

Barton Wright, *Pueblo Shields* (Flagstaff, 1976).

J. Wyatt, "Some accounts of the manners and superstitions of the Adelaide and Encounter Bay tribes," in J. Woods ed., *The Native Tribes of South Australia* (Adelaide, 1879), pp. 157-181.

J. Young, *The Prose Edda* (Berkeley, 1954).

R. Zingg, *The Huichols: Primitive Artists* (New York, 1938).

www.ingramcontent.com/pod-product-compliance
Lightning Source LLC
Chambersburg PA
CBHW081827170426
43202CB00019B/2973